The Thinking Executive's Guide to Sustainability

The Thinking Executive's Guide to Sustainability

Kerul Kassel

First published in 2014 by
Business Expert Press, LLC
222 East 46th Street, New York, NY 10017
www.businessexpertpress.com

ISBN-13: 978-1-60649-419-6 (paperback)
ISBN-13: 978-1-60649-420-2 (e-book)

Business Expert Press Environmental and Social Sustainability for
Business Advantage Collection

Collection ISSN: 2327-333X (print)
Collection ISSN: 2327-3348 (electronic)

Cover and interior design by Exeter Premedia Services Private Ltd.,
Chennai, India

First edition: 2014

10 9 8 7 6 5 4 3 2 1

Printed in the United States of America.

Abstract

We live in an increasingly global economy in which the effects of shrunken economies, broadened communication, and widespread meteorological incidents associated with climate change are leaving virtually no one untouched. As a result, a working knowledge of concepts such as the triple bottom line and sustainability have become mandatory. Systems thinking, foundational for grasping these concepts, is based on transdisciplinary theories, deriving in part from biology, physics, economics, philosophy, computer science, engineering, geography, and other sciences. Specifically it is the study of systems, including all life-forms, climate phenomena, and even in human learning and organizational processes, that regulate themselves through feedback.

The media and the public have become savvy to corporate greenwashing, and government regulation, already pervasive in Europe, is imminent in the United States of America. Business practices are a subsystem of human activity, which is itself a subsystem of the biosphere we all depend upon for services, such as clean air and water, sufficient soils to produce food, and moderate weather. Corporate sustainability practices are in the midst of becoming a required aspect of the social license to conduct business, and the use of a systems framework provides a coherent and eminently sensible way to comprehend the structure and logic that underlies this transition.

Green business efforts and stakeholder initiatives undertaken by those without the requisite understanding of sustainability and the trends related to it in the world of commerce risk adverse press, activist pressure, regulatory constraint, added expense, reduced revenue, and lowered valuation.

This book offers a practical, relevant, and easily grasped overview of sustainability issues and the systems logic that informs them, supported by empirical research and applied to corporate rationales, decision making, and business processes. Intended for business professionals seeking concise, reliable, and current knowledge and trends, it will support them in leading their organizations' corporate sustainability, social responsibility, and citizenship efforts so that they can remain competitive and successful.

Keywords

corporate sustainability, corporate social responsibility, corporate citizenship, stakeholder management, sustainable business, green business practices, metrics options

Contents

What Readers Say About

The Thinking Executive's Guide to Sustainability

This work touches on a number of concerns that many firms are still searching for answers on. It's a great read to get senior managers and corporate governance folks thinking…as they ultimately hold responsibility for the true long-term impact of their business practices on society.

—Brock Nicholas, MBA, Senior Vice President,
Harmony Development Company

Kerul and her co-authors are very broad, and at the same time focused thinkers. They challenge assumptions that make you think differently and better (see Chapter 3 where she challenges different myths). Any business professional can use and benefit from the clarity, directness, and relatedness of this book's important and timely content.

—Jon London, President, The Improved Performance Group

A brilliant piece of work by Kassel and her co-authors. If you care about the future of our world—now is the time for us to stand and take action. It is incumbent upon each of us to understand the critical issues we face. Kassel and her colleagues explain the social and economic consequences of our continued ignorance while showing us the path to a brighter future. Take heed!

—Bernadette T. Vadurro, Author, Business Speaker,
and President, Speakers Live, Inc.

This book provides companies with new ways to improve sustainability in small and large ways, and how to measure these impacts and their resulting financial benefits. It's a must read for any company looking to positively impact the environment, the communities it touches, and its own bottom line.

—Beth Maxim, Manager, Global Learning Operations,
Tyco International

Readers will just sail through this well-written, timely, easy-to-understand book on a complex and important subject.

—Ruth Ann Harnisch, President, The Harnisch Foundation

This book is a wonderful example of advocating without preaching. Kassel and her colleagues provide succinct and powerful evidence that companies need to seriously consider changing their thinking on sustainability, not just because it would be "doing good" but also because it would be best for their organization's bottom line. It is extremely well written and respects the fact that its intended audience has very little disposable time. Although each chapter builds nicely upon the one before, this book could and should serve as a resource for executive decision makers in companies large and small. I predict that when leaders begin to read this work they will want to keep going and will want to share its ideas with those they lead. I would urge anyone who is serious about doing well and doing good to pick up this book. It is a somewhat rare example of a work that is at once packed with important information and is also a pleasure to read.

—Chuck Berke, Executive Coach/Principal, Berke Associates, LLC

A surprisingly comfortable read for a scholarly text that engages such a vital and complex topic, the author offers business professionals at all levels a conceptual as well as a practical blueprint for how to think and act in ways that just might make future environmental activists unnecessary!

—Four Arrows, aka Don Trent Jacobs, PhD, EdD,
Fielding Graduate University faculty and author of
Unlearning the Language of Conquest,
Differing Worldviews, and *Teaching Truly*

The sustainability theme makes sense as to why we have specific sustainability programs in place within corporate America. Having an ecologically minded society cannot end with individuals. We need our various public, private, and government sectors involved with sustainability programs as an investment into our future. This is a great book for our corporate citizens to start understanding the sustainability concept and its applicability.

—Carl G. Fsadni, CEO - Director,
Oak Ridge Group LLC and MALTA.net

Introduction

Don't be put off by the inclusion of "systems thinking" in the description of this book. Systems thinking isn't some complex, intangible notion that involves a lot of science and math. It is a way of better understanding the world based on tangible, observable facts that we live amongst every single day, at work, at home, within our own bodies, and scaling back out to our communities, nations, and economic and industrial systems.

The aim of this book is to introduce people in business, at any level and in any industry, who have repeatedly heard the terms "sustainability," "triple bottom line," or "corporate social responsibility" in a business context but want to learn what it means, as well as whether and how to do it.

For those already familiar and conversant with the terms, but not with systems thinking, it will add an understanding of how to pursue sustainable processes and practices. This book offers a succinct and easy-to-assimilate overview of these topics and uses the most recent information in making the business case for integrating such processes and practices into any firm, and for doing it more deeply. This may be a book to offer to colleagues who don't yet understand growing opportunities and pressures, to plan for and address issues related to incorporating the planet and people to secure and enhance the prosperity orientation of commerce.

Financial success is judged on a short-term time frame. Quarterly performance, annual gains, and exponential growth are rewarded but, as will be discussed in pages that follow, emphasis on rapid and explosive growth in the near future frequently has had a negative impact when viewed over an extended span of time. If businesses wish to remain competitive and serve their shareholders, they will be required to contribute to social well-being in the long term. As incentives shape behavior, structuring them to make an impact offers more leverage than continual focus on the back end.

The seven chapters that follow are designed as a logical and easy-to-follow flow of information, supported by a host of documented and up-to-date facts. Each chapter dives progressively and gradually deeper into the concepts and ideas, and each is concluded with a brief summary.

There is, hopefully, a balance of enough repetition in each chapter so that each of them stands on its own yet in total brings home the most important points without being annoyingly redundant. A number of additional resources are offered in the endnotes for those who want to delve further into some of the areas covered.

The authors' hope is that readers find the information in this book useful, illuminating, and compelling, so that it serves as a springboard for further efforts for organizations to better themselves, finding ever more effective ways of innovating into the evolving 21st century model of business.

Acknowledgments

Thanks, first, to David Parker, for suggesting this book project to me, and to Cindy Durand and Destiny Hadley for helping to herd it through the production process, all of Business Expert Press. Gratitude goes out to my colleagues and reviewers, who are too many to name, but particularly to Nettie Bartel, Beth Maxim, Brock Nicholas, Chuck Berke, Romi Goldsmith, Charlene Hamiwka, Pam Rowe, and Karen Bogart. Their warm encouragement and stronger shoulders (leaned on heavily and often), as well as their gracious and honest feedback, helped shape this work.

I would especially like to thank my contributing authors, Kim Hedberg and William Paddock, for all of their hard work and diligence.

Kim, a business consultant working with companies on sustainability issues, professional development, business plan writing, and financial analysis, helped with the chapter on the phases of commerce, a number of informative diagrams, and the initial editing process. With advanced degrees in both Hydrology and Business, a B.A. in economics, and her skills in teaching, environmental consulting, and business consulting, she has a sharp ear for an effective and persuasive approach.

William is the founder and managing director of WAP Sustainability Consulting, a Nashville based sustainability-consulting firm to Fortune 500 companies and leading manufacturers. His experience establishing, building, and leading sustainability at the enterprise level span more than a decade. William's expertise in the nuts and bolts of sustainability metrics contributed to significant parts of the last three chapters.

The assistance and input of these deeply knowledgable contributing authors was invaluable. Appreciation goes to the International Society of Sustainability Professionals, who made possible a connection to be forged between Kim and me, and to Toby Thaler for his thoughtful referral of William.

Quick Reference Glossary and Acronyms

Anthropogenic: effects that are caused by the activities of humankind.

CDP—Carbon Disclosure Project: the original name of a non-profit organization, now known as CDP, offering a system for companies and cities to measure, disclose, manage, and share vital environmental information.

CSR—Corporate social responsibility: the creation of shared value through environmental restoration, stakeholder engagement, appropriate governance, and economic prosperity.

ESG—environmental, social, and governance issues.

GHG—a gas emitted into the earth's atmosphere that contributes to global climatic change, known as the greenhouse effect, by absorbing infrared radiation. The gases primarily responsible include carbon dioxide, methane, nitrous oxide, and ozone.

GRI—Global Reporting Initiative: a non-profit organization that promotes sustainability through a framework designed to assess and report sustainable organizational practices.

LCA—Lifecycle assessment: a study that compiles and evaluates all of the inputs, outputs, and the potential environmental impacts of a product throughout its lifecycle, from extraction, refining, and manufacturing through distribution, use, and disposal.

System: a set of parts, people, atoms, or anything interconnected in a way so as to produce its own set of internal relationships, patterns, and other dynamics. A system is generated, driven, constrained, or pressured by inside and outside influences to which it responds in its own characteristic way.

Sustainability: meeting the needs of the present without compromising the ability of future generations to meet their own needs.

TBL—Triple bottom line: an accounting framework designed to integrate ecological and social performance into financial reporting, and intended to include natural and human capital to measure organizational success.

Chapter Synopses

Chapter 1

- The business world, as well as society, has reached a tipping point: economic, social, and environmental incidents have forced industry to begin to address how they conduct business, as evidenced by the tremendous growth in sustainability practices and reporting by businesses of all sizes.
- Research on the cause of the problems now being faced reveals the need to develop an updated worldview, one that is more adequate in explaining the reason behind those problems.
- Studies have shown that this revised paradigm, based on a principle of interdependency between systems, has gained traction, not only in science, but also in many disciplines, and in commerce.
- In the broader view, sustainability is the ability for economic, social, and natural environmental systems to coexist in a dynamic equilibrium far into the future, the capacity to meet current needs while ensuring the ability to meet the needs of future generations.
- Technical solutions developed using an outmoded perspective will necessarily fail, as they will continue to result in the same kind of issues that currently exist.
- A system is a self-organizing entity that generates inflows and outflows. Within the system are hierarchical, differentiated subsystems which act in interdependent relationship. Central control is balanced with subsystem control, providing resilience.
- Subsystems that are misaligned with the purposes of the overarching system create imbalances, resulting in destabilization and collapse. Increasing size creates increasing delay in observable change.

- These systemic qualities are present in all business and industry, just as they are in all biological systems. As business has evolved from highly local tribal bartering to immense and widely distributed transnational commerce, human thinking is in the process of evolving from a mechanistic and reductive worldview to one informed by these principles.
- The ability to think in terms of patterns and relationships rather than simply causes and parts in isolation, is becoming a critical business skill.

Chapter 2

- Most experts agree that any definition and application of sustainability requires attending to a balance of economic, social, and environmental concerns, in both present and future contexts.
- This triad of components underlies the foundation of triple bottom line thinking and accounting practices, as well as most sustainability frameworks.
- Sustainability is a systemic concept with an expanded time frame, incorporating an understanding of how civilization has reached this point and endeavoring to create parity for the current as well as future inhabitants of the planet.
- Corporate social responsibility and sustainability efforts are currently voluntary from a legal perspective, but are becoming part of a social expectation.
- As impacts and issues vary between industries and firms, there is no single best way to "do sustainability." Programs and initiatives run from the substantive, wide-ranging, and deeply integrated to shallow greenwashing.
- A systemic approach requires awareness of and interaction with stakeholders, those parties impacted by the firm, positively, negatively, or both, such as employees, vendors, nongovernmental organizations, local communities, government, customers, and distributors.

- A number of pressure points are driving responsible business practices, including public sentiment, media exposure, activist campaigns, government regulation, institutional investors, and insurers.
- The good news is that there is a solid business case supporting those practices: cost savings, managed risks, reputational enhancement, customer loyalty, employee engagement, and new revenue streams and markets are a few of the payoffs.

Chapter 3

- A number of current economic assumptions are flawed, such as
  - that growth benefits everyone;
  - that people make decisions rationally and only for their own benefit;
  - that infinitely continual growth is possible;
  - that costs for business impacts not directly incurred by a firm are outside its responsibility.
- Research has found that
  - growth benefits few, and that it fuels growing income inequity;
  - emotion drives decision making as much or more than logic;
  - cooperation is as potent a force as competition;
  - constrained resources are forcing disruptive innovation in business;
  - commercial interests can no longer externalize their impacts on society and the natural environment.
- Commonly held assumptions about corporate social responsibility and sustainability in the business context are that
  - it's too expensive;
  - too intangible;
  - not important enough to pursue.
- Instead, sustainability efforts can and often do
  - lower expenses;
  - are tied to solid metrics;
  - will boost any short- and long-term strategic plans.

- Industry is among the most significant drivers of the degradation of air, water, soil, oceans, climate, and biodiversity, all vital natural capital at considerable risk of becoming ever more scarce, expensive, and insecure. These problems offer solutions for enterprises to address.
- A systemic examination of the social side of sustainability reveals that economic and environmental issues are related to issues of poverty.
- Access to clean water, food, energy, education, healthcare, and livelihoods are all areas around which enterprises can build new businesses and models.

Chapter 4

- There is significant overlap between extraction, production, distribution, consumption, and disposal.
- Sustainable products include those that utilize fewer resources both in their manufacture and use, that last longer, that offer improved performance, and that are designed to be recycled or reused instead of landfilled where further impacts (air, water, soil pollution, and the emission of methane gas) can occur.
- Sustainability in commerce includes integrating such considerations into the design, sourcing, manufacturing, and distribution processes along with practices that reduce negative impacts, enhance positive social, environmental, and economic impacts, or both.
- Systemic examination throughout the entire process of commerce reveals opportunities to make more with less.
- The systems perspective requires that consequences and impacts are determined at each phase—including environmental (air, water, land, and species impacts), social (worker, community, and stakeholder interests), and economic.
- Tools such as industry process maps have been developed. Simple systems maps can help guide the management team to analyze all impacts, both positive and negative, for each step

in the development and delivery of a consumable product or service.

- Process and system mapping can be accomplished in-house or developed through consultants.

Chapter 5

- Industrialization, made possible by scientific discovery and the use of fossil fuels, made rapid population growth possible, stimulating further industrialization and resource use.
- Slowly, beginning a century and a half ago, and much more rapidly in the last four decades, has come a realization that industrialization, population, and technology are destabilizing forces on the biosphere.
- Responsibility to restore the planet's dynamic equilibrium is shared among 7 billion individual humans (with those in emerging nations using a fraction of the resources of those in advanced nations), non-profits, government, and for-profit business.
- Industry, however, is both the biggest driver of destabilization and the most capable player in taking the lead. It is the sector with the speed, knowledge, expertise, influence, incentive, innovation, result orientation, access to capital, and collaboration potential flexibility, and innovation to solve the big problems facing humankind.
- The benefits to industry for doing so are many, including conservation of and better access to capital, long-term growth, expense reduction, employee engagement, brand recognition, and customer loyalty, to name a few.
- With the advent of hybrid and B corporations, enterprises may be adjusting to the need to balance profit primacy with environmental health and social well-being.

Chapter 6

- The pursuit of sustainability is not a one-size-fits-all proposition; it must be tailored to each organization according to its size, industry, values, and capacities.
- Some of the more common resources for measuring and implementing environmental impact and change strategies include inventorying greenhouse gases, conducting lifecycle analyses, or utilizing a standard reporting framework such as that offered through the International Organization for Standardization and CDP's programs.
- Greenhouse gas inventorying is a form of accounting that measures direct and indirect emissions, and can identify areas for improving efficiency and reducing impacts.
- Lifecycle assessments follow a product from cradle to grave—from the extraction of raw materials through refinement, product manufacture, transport, consumption, and disposal. There are both environmental and social lifecycle assessment tools.
- CDP offers programs for measuring, managing, disclosing, and sharing an organization's risks related to the environment, particularly climate, water, and forest-related issues.
- The Global Reporting Initiative offers a framework that integrates not only environmental issues, but also social, economic, and governance components as well.
- While full use of these tools, standards, and frameworks are the province of larger organizations, they offer smaller and mid-sized firms plenty of guidance to get started with metrics.
- The use of any of these frameworks and tools refocuses measurement to align with a more systemic understanding of the interface of industry with the larger systems of society and environment in which it operates.

Chapter 7

- Resources needed to begin working sustainable practices into an organization include familiarity with and an ability to discuss the business case for it and how it is aligned with an organization's strategic plan. Support from top leadership, including championing efforts and supporting with human and financial resources are key, as are mental toughness, persistence, and patience.
- Sustainability efforts occur in a cycle of planning, baselining, goal setting, measuring and monitoring, reporting to stakeholders, and starting another cycle by reviewing and updating goals.
- Some leverage points have more power than others. Changing mindset is the most potent, followed by changing the objectives of the organization, changing the rules and incentives of the organization, and improving access to information.
- Any plan will need to address resistance, which takes many forms, including aversion to change, short-term thinking, mental and strategic shortcuts, ignoring empirical facts, and conflating wants into needs.
- A focus on only low-hanging fruit is likely to result in minimal ad hoc projects that don't address an organization's material impacts.
- Measurement can have a number of components and should address elements of governance, environmental, and social indicators. There are a number of principals associated with these indicators, such as relevance, completeness, consistency, transparency, accuracy, quality, balance, clarity, reliability, and boundary that guide reporting.
- The longer society delays on addressing environmental, social, and governance issues, the more severe will be the consequences. The future of industry, and of society, may be constrained by necessity. It will take the concerted and combined attention of individuals, communities, development and research organizations, universities, governments, and enterprises of all kinds in pursuing a sustainable future.

CHAPTER 1

Systems Simplified

A Tipping Point?

Business trends provide essential information that business leaders ignore at their firms' peril. One such trend is the integration of environmental, social, and governance (ESG) issues into firm operations. A visible example of this trend is sustainability reporting, which has been growing for decades and continues to expand in spite of years of global recession. While reporting is no longer a new practice for many firms, a joint study released in 2013 indicated that calls for transparency and reporting were resulting in increasing mandatory government or market regulation in a number of countries, prompting firms to sharpen or extend their efforts, and their reports.[1]

Now that reporting, and the initiatives that accompany them, have become expected, firms are seeking and finding means to turn the application of sustainable practices toward tangible and intangible value creation, as well as competitive advantage, and not a little risk management. Firms have only more recently begun to develop a full understanding of the benefits, and of the risk components. A 2013 survey by Ernst & Young noted that most firms were not addressing material sustainability-related financial risks.[2]

As business trends go, sustainability has ramped up in the past few years. A survey of senior business executives by KPMG conducted back in 2010 found that 62% of the firms responding had a sustainability strategy in place, representing a 25% increase from 2 years earlier.[3] More than half of these public and private companies had either already issued a public report or were planning to do so in the near future. Executives participating in the survey cited potential cost reductions as a key driver, with risk management, brand enhancement, and regulatory requirements as additional rationales. A follow-up report, KPMG's International Survey of Corporate Responsibility (CR) Reporting, stated, "CR reporting has become the de facto law for business"[4] and that "of the 250 largest global companies, fully 95% now report on their CR activities."[5]

According to the surveys, sustainability is increasingly viewed as a strategic focus for both new business opportunities and innovation in processes, practices, and systems. This sentiment was similarly expressed by the respondents of a survey published in 2012 by a joint effort of Ernst & Young and GreenBiz Group.[6] On the minds of the leaders of these publicly traded organizations were entry into prestigious sustainability stock indices and high-profile rankings. Such ratings and rankings benefit a firm, making it worthwhile to disclose their practices, as the call for accountability grows through the media, non-governmental organizations (NGOs), shareholder activists, and customers. Fully two-thirds of firms responding to this survey reported an increase in inquiries from shareholders and investors about sustainability-related issues. A June 2012 article on Forbes.com suggested that sustainability reporting is hitting a tipping point, "as investors, governments and even many influential corporations come to see such disclosure as a key mechanism for strengthening markets and essential to building a sustainable economy."[7]

That is not to say that such reporting is not without its challenges. There are some direct costs associated with baselining, implementing programs and monitoring them, ongoing measurement, and reporting. Uncertainty about trends in the regulatory environment adds risks, as does customer sentiment and willingness to pay for added costs associated with sustainability efforts. There is no guarantee as to whether those efforts will be well received by consumers, the media, and shareholders, as well. The growing practice of reporting, though, is evidence that it pays to make the effort, in spite of the risk. The business world is on the brink of understanding itself as a system that exists in relation to and is interdependent upon other systems: the economy, the government, the public, and the natural world.

A development in the insurance industry offers another example that illustrates the point. In the Market Trends section of their website, the International Insurance Society reported on a United Nations initiative called Principles of Sustainable Insurance, which the insurance trade group launched in June 2012 at their annual meeting. The International Insurance Society describes the commission responsible for these principles as "a group of leading insurance industry institutions… committed to embedding in industry decision-making environmental, social and governance (ESG) issues relevant to the insurance business, and, more broadly, to sustainable development, through implementation of" these principles.[8] The effort is intended to require that institutional investors account for ESG issues as part of their fiduciary obligations to invest in the best interests of their beneficiaries. Taking stock of the losses incurred through management mistakes and misdeeds, public relations crises, and environmental disasters, the insurance industry has started to recognize the complexity of interactions among the players, or subsystems, that increase their exposure.

Boards of directors are waking up to the recognition that their fiduciary responsibilities of making organizational decisions and guiding management behavior include generating and maintaining the trust of the public. Increasingly, the public requires transparency as a requirement to elicit that trust.

Systems and Sustainability: What This Book is About

There is certainly a trend afoot, one that appears not to be a fad, but has been building up over the last few decades. Most people have heard of the term "green," and while some in management may consider it a "flavor of the month," the developments described above are evidence of not only the staying power of the trend, but its increasing reach and depth in business practices. This book offers an overview of these developments, an explanation of why they have built such steam, and a way of incorporating the forces behind that explanation into an enhanced business model that offers better results.

There is a growing argument that business as usual isn't cutting it anymore. Examining that system and seeking to understand how and why the system has been working (or breaking) allows firms to apply more than just a band-aid. The aim of this book is to give readers a more robust grasp of the forces that influence their firm, industry, economy, society, and beyond, for better and for worse, so that they can make decisions that offer many competitive advantages.

There are underlying assumptions embedded in the perspective offered in the next seven chapters. One of them is that the authors accept established scientific theory, including evolution. An extension of this is that all human activity, including the business world, is influenced by the same forces that created the natural world with all its life-forms, including *homo sapiens*. A second assumption is that a systemic mindset, or paradigm, is an extremely helpful add-on to the current dominant mindset. The latter mindset might be called "mechanistic," meaning that the thought process used to explain the world is compared to how parts of a machine work together. Another characteristic of the current dominant mindset is its "reductionist" aspect, meaning an approach to understanding complex things by reducing them to their parts, the whole being not more than the sum of those parts. Systems thinking, while incorporating some aspects of the mechanistic and reductionist models, goes beyond to offer a more realistic, logical, and sufficient explanation of the how things work. See Table 1.1 for some distinctions between mental models.

That is not to suggest that a systems paradigm explains everything and that it is the ultimate mental paradigm, but rather that it is the next evolution in the understanding of the world. Human thinking evolves. One

Table 1.1. Comparison of Mental Models

Current dominant mindset focus	Systems mindset focus
Unrelated parts: parts as separate from and independent of the whole or larger system.	Relationships: interaction and interdependency of parts between themselves and the whole.
Structure: the organization of the parts, and their traits. The map.	Process: how the parts, inputs and outputs, and internal and external forces create change. The territory.
Contents: ingredients, parts, boundaries, factors, variables. Symptoms.	Patterns: changes in quality and quantity, influenced at leverage points. The source of symptoms.
The parts: focus on the parts individually, comprising a total that is no more than the sum of the parts.	The whole: focus on the whole, and how the parts relate to each other and synergistically combine to create a whole that is more than the sum of its parts.
Linear movement: one way, one direction. Cause and effect.	Multidirectional movement: many ways, many directions. Complexity.

example is the transition from a belief that the world was flat and the sun revolved around it to viewing the earth as round, revolving around the sun, and beyond, to the capability of landing vehicles on other planets. Humankind is in the process of shifting their paradigm from viewing the aspects of the world as separate parts to seeing them as intertwined in relationship with each other.

As a simple example, non-compliance with company policy may appear to be a problem originating with poor supervision, so compliance targets are set, to which incentives are attached. Yet the problem may have its source in insufficient training, a discrepancy between firm culture and policy, a prioritization of key performance indicators that conflict with compliance, inadequate communication between departments, or other sources elsewhere within, or outside, the organization.

The first three chapters of this book provide a simple, clear explanation of the systems framework. This framework, informed by business news, scientific research, and other developments, necessarily leads to concerns about sustainability. These chapters set the stage for applying the systems framework to the world of commerce and management decision making covered in Chapters 4 through 6. Chapter 7 offers a brief overview of ideas with which to get started in your business and a discussion of where these developments may be headed next, with examples of emerging business and economic models, in preparation for the next wave of change. Each chapter goes successively deeper into thinking systemically and applying that thinking toward sustainable business solutions.

What Is Sustainability?

The concept of systems has been introduced, but what is meant by sustainability? Is it the same thing as corporate social responsibility (CSR), or corporate citizenship? Does environmental health and safety cover it? Is "green" the same thing? The answer to these last three questions is no. Sustainability is an overarching term, a concept that is contested (see Chapter 2), misunderstood, misused, and abused, but at the core it simply means the ability for economic, social, and natural environmental systems to coexist in a dynamic equilibrium far into the future. The concept of sustainability inherently takes a long-term view. Human impacts upon the planet, or "biosphere," should not cause scarcity in key resources such as drinkable water, breathable air, sufficient fertile soil, and stable, productive oceans. Sustainability also incorporates overall human well-being, including economic systems, in that they must be adequate to avoid widespread social suffering and upheaval.

While social aspects of sustainability must be attended to as intentionally as environmental, the state of natural systems may suggest more urgency. Increasingly, scientists of all disciplines are reporting that our behavior is pushing the planet's dynamic equilibrium to the brink, with growing damage to natural systems, and leading to social impacts. Massive losses of arable soil due to over-tilling and erosion, fresh water use far outpacing the ability of aquifers to replenish it, overfished oceans that are becoming more acidic and salinized, and air quality increasingly imperiled in many locations are some of the environmental impacts. Nature can take care of itself to a point but it is likely to take thousands, if not tens or hundreds of thousands of years, particularly in regard to biodiversity, to recover only some of the losses the biosphere has incurred in the last couple of centuries. Much of the damage has occurred only in the last 50 years. If the previous decade is any indication, nature's rebalancing efforts may continue to wreak havoc in the meantime. Climatologists predict increased extreme weather patterns, including record-setting heat, higher incidence of wildfires, drought in some widespread areas and floods in others, and extensive loss of crops due to these and related other problems.[9]

For those who tend to believe that technology is the answer, while society can hope to find viable technical solutions, it would be a mistake to implement fixes that resolve one problem while causing another. There is no lack of double-edged swords that have provided short-term solutions while inflicting inherent long-term costs. The Green Revolution increased agricultural production in part by using dichloro-diphenyl-trichloroethane (DDT), a pesticide from which ecosystems are still recovering. Shifting agricultural production to bio-fuels has altered agricultural production and increased food prices around the world. Hydrofracking (also known as fracking), while shifting away from problematic oil drilling to cleaner natural gas, is increasingly being shown to endanger water supplies.[10] Carbon sequestration may be less effective and more dangerous than anticipated, with leaks to the surface and possible induced seismic activity.[11] Insect and herbicide resistant crops are beginning to backfire, leading to herbicide resistant superweeds and contamination of conventional and organic crops.[12] The use of aerosols in refrigerants and propellants resulted in a significant gap in the protective atmospheric layer, and the pervasive use of fossil fuels is altering our planet in a number of harmful ways, threatening not only ecosystems, but also our own species. While any remedy has its downside, shortsightedness will only be more expensive and harder to fix later.

You may be inspired by the potential opportunities, or perhaps you see the writing on the wall, that customers, employees, the media, and government regulation are shifting the ground beneath the feet of industry. If you want to know how commerce can contribute toward sustainability, continue reading.

There are many books that discuss sustainability initiatives in business, and this book stands on the shoulders of some giants in this field, notably Peter Senge,[13,14] Paul Hawken,[15] Amory and Hunter Lovins,[16] Donella Meadows,[17] Peter Checkland,[18] and Russell Ackoff[19]—as well as Wayne Visser's excellent set of summaries in *The Top 50 Sustainability Books.*[20]

While these notable authors and others have detailed a systemic approach to sustainability in a business context, there are few books that offer a concise and very accessible overview that incorporates both the thinking that underlies the premise of sustainability and its application to business practices. Without the systemic foundation, almost all efforts at creating sustainability will fall short. The momentum toward sustainability in business practices continues to gear up, to which the documentation of many recent developments and events included in the following chapters attests. It is inevitable and irrevocable, and those leaders who are not equipped with the knowledge, competencies, and tools to help them navigate through this uncharted territory will lose advantage and opportunity. In a brief format, this book is an easy-to-use guide that boils down the big picture sweep of a complex topic, supported with up-to-date documented facts.

The following chapters are an overview of sustainability in commerce from a simple systems perspective. It is not primarily about the organizational dynamics and values that drive firm activities and policies, but which are still necessary for successful commercial initiatives grounded in a sustainable sensibility, although we touch on this subject in Chapter 7. The focus of this book is to offer a different, more comprehensive, way to understand the changing forces that are making business so complex today, and how to harness these forces toward successful business results.

Leaders may sometimes find it hard to reconcile these pressures, and feel pulled in different directions in serving the interests of shareholders, employees, vendors, customers, the media, financial analysts, the government, local communities, and the environment. While it may not appear to be so, these groups and their concerns are moving in the same direction, though not at the same rate and not necessarily in a linear path. The goal of this book is to help you become adept at identifying the existing patterns in the system, to identify leverage points at which to exert pressure to create the desired change.

Business Practices Viewed from a Systems Perspective

In case you are not yet convinced that a systems mindset applies to business concerns, that they are qualitatively different than the "natural" world because they are a human construct or for some other reason, consider the following points.

From an evolutionary perspective, the structure of a commercial enterprise replicates natural systems on a number of levels: on an individual business level as a social unit, and as an entity, and on a bigger scale as a component in a trade system, and the trade system as a component of not only the human system, but of the global biospheric system (see Figure 1.1).

As an entity, even if it is a conceptual one, a business starts its life as **self-organizing**, comprising a few individuals with an idea for a product or service. These few people perform the different tasks necessary to **generate inflows and outflows** of **stock** (materials, parts, products, services, and money being the more obvious flows). As the business expands it becomes more complex and **hierarchical**, and departments (**differentiated subsystems**) evolve and then grow, not only in size but also in the depth and breadth needed to handle an increased and wider set of functions more **efficiently**.

Economic viability	Social equity	Environmental health
Risk management-reputation market share-regulation technological innovation-profit asset management-fair trade	Health safety/security-human rights social justice-well- being-choice cultural diversity and preservation	Biodiversity-climate change air quality-water quality/supply habital preservation-soil viability consumption of natural resources

SYSTEMS

LONG RANGE

REDUCTIONIST

SHORT RANGE

Biosphere

Future generations

Developing world

Developed world

Industry

Organization

Individual

BIOCENTRIC

ANTHROCENTRIC

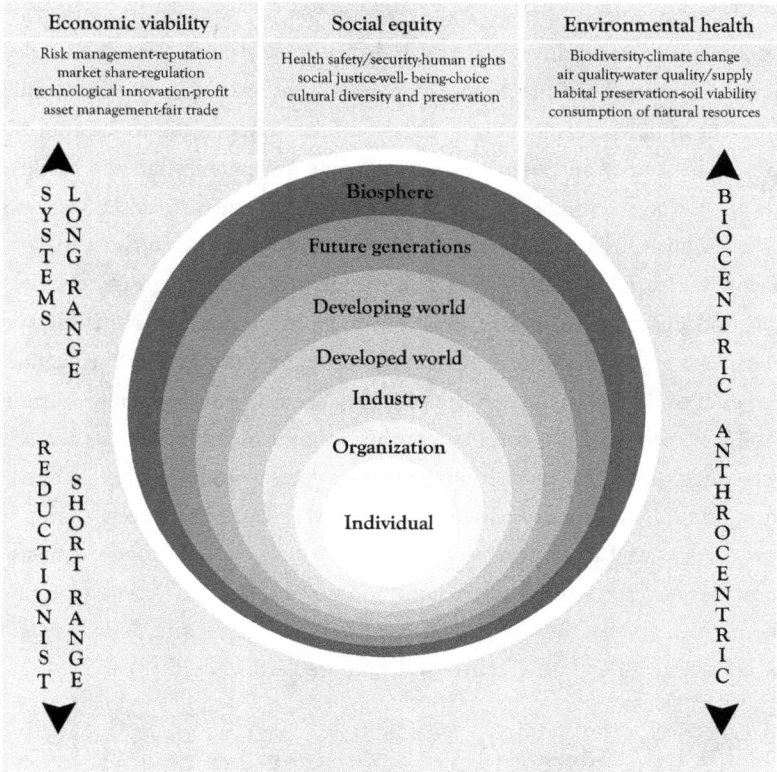

Figure 1.1. *Simplified sustainability framework illustrating systemic interconnections.*

These departments are in **interdependent relationship** with each other, in which **central control is balanced with departmental control**. This gives the business greater **resilience** to withstand negative impacts to any department. **Relationships are stronger within** these departments than the relationships the departments have with each other.

Feedback loops, such as slow sales, returns, parts availability, and distribution issues, are the information mechanism by which business determines a need to change behavior. The behavior change results in a **dynamic equilibrium** where the business regulates itself to adjust to constraints, delays, and surprises. **With increasing size comes increasing delay** for the organization to recognize problems and act, and for the actions to have an effect.

Interdependence exists not only within the organization, but also with vendors, customers, investors, and industry trade groups. The term stakeholder, explained in more depth in Chapter 2, has its roots in systemic thinking. The interdependence continues beyond the human realm to the available physical resources and energy needed to house the organization and produce its goods and services. The links are beyond such immediate and direct ones as computers, telephone, and paper, but extend further to software makers, raw materials for ink, suppliers for all the bits and pieces of computers, servers, and peripherals, cross-national telephone networks and switching systems, energy and electricity for manufacturing, not to mention the facilities for each of these supplies and the land they occupy, and their supply chains. The interconnectedness continues farther, as none of these products and services are possible without all the people working to mine, refine, manufacture, administer, deliver, and service these further links. These people must be included, as well as the transport that gets them to work, the food they eat to provide energy to work, and the communities and families that they depend upon. And none of those would be viable without secure, safe, and affordable energy, clean air and water, fertile soil, and predictable weather patterns.

Why Understanding the Systems Framework Is Essential for Understanding Sustainability

The social world is increasingly interconnected, not simply through social networking but in terms of resource extraction, supply chains, labor, and markets. The globalizing human community of commerce is a reflection of the physical planet that all human activity depends upon for countless needs. Understanding how industries, and the biosphere, are systemically (not just systematically) coupled provides insight into trends and opportunities, as well as risks and exposures. Such an awareness has become necessary, not only to conduct legal, profitable business, but also to maintain a "social license," meaning the legitimacy in the eyes of the public, media, and the government, to do business in the 21st century.

The Limits To Growth first popularized the concept of "systems" as a model for understanding the world.[21] One of the authors, Donella Meadows, went on to write a primer on the systems model, defining a system as "A set of elements or parts that is coherently organized and interconnected in a pattern or structure that produces a characteristic set of behaviors, often classified as its 'function' or 'purpose'."[22] As a supply chain, a manufacturing scheme, a set of distribution channels, or an organization's information technology is a system, so is the organization, its departments, and each of the individuals within it. A system has identifiable parts, some tangible, some intangible, that are interdependent and interconnected, and together act differently than the parts do alone.

Systems have the capacity to regulate themselves based on feedback or signals. In an organization, as widget sales increase, widget inventory is reduced, which signals the production department to create more inventory, and the purchasing department to order more widget parts. There are flows of both physical parts and information.

To develop more of a systemic thinking model, let's explore an example of a (sub)system, a jet airplane (see Figure 1.2).

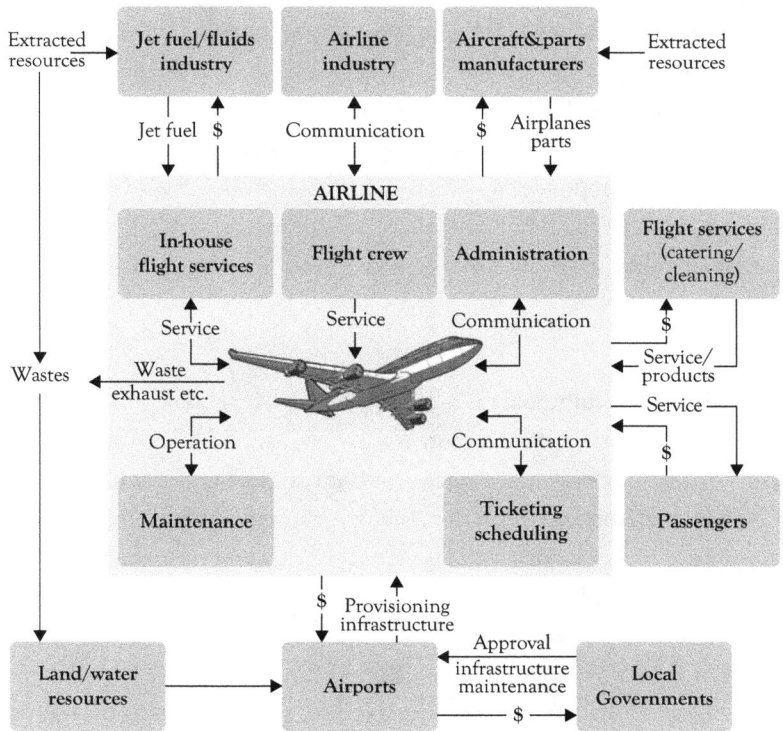

Figure 1.2. Simple systems diagram of using an airplane for illustration. Arrows are flows, and stocks are what flow along the arrows.

Elements

A jet airplane is a system of **elements** or parts, such as engines, a cockpit with instrument panels, control yokes for steering, a body with wings, rudders, wheels, tires, and a flight cabin, all interconnected in a structure for the **purpose** of moving people from one location to another. The airplane itself contains a number of subsystems that specifically support its purpose: a system for acceleration, another for braking, a third for electrical, another for steering, and so on. These subsystems often influence each other through **flows** of **stock**: air provides lift, the oil lubricates the engine, batteries provide power to start the engine, and the engine provides the power to recharge the battery, while jet fuel powers the engine. Stocks, as will later be explained, can be material, energy, or information.

The airplane is made up of a number of materials, including various types of metals, plastics, resins, rubber, and glass; you could substitute some elements for others of similar type, without too much impact on the functionality of the vehicle. Leather or a synthetic could replace fabric for seats; newer, lighter metal alloys and high impact plastics have replaced prior technologies, for example. Take away some parts, however, like the engines or the wings, and the airplane ceases to perform its function. At that point it is no longer a system and it ceases to function as designed, just as a business could not operate without products, services, or customers, or an animal could not function without a heart or a brain.

Interconnectedness

A jet airplane is commonly understood to be a single object with unambiguous boundaries. There is more to it, of course, and without consideration for these aspects, the airplane will not be able to serve its purpose. Looking at the airplane from the systems perspective allows us to consider it within a bigger picture, the **dynamics** of how the airplane influences and is influenced by other objects, materials, factors, and forces (which are actually elements and systems). The focus is on the **relationships** between them.

The most obvious way the airplane is a subsystem, part of a larger system upon which it is interdependent, is that it needs jet fuel, with its own industry of extraction, refining, distribution, and sales. In order to fulfill its function it also needs passengers, for whom the airplane must be safe, affordable, and convenient enough to use. The route the airplane travels, therefore, is dependent on where those passengers need to go. For now, the need to understand the demographics of the passengers will be skipped, but suffice it to say that without people for whom the airplane is the transport of choice, it would not serve its purpose. Another component of the larger system within which the airplane operates is the airports to which it travels and that must be regularly maintained. The airplane also requires a labor force to keep the vehicle fueled and maintained (another set of industries for parts and mechanic training), as well as driven and paid for. Furthermore, it is usually one of a number of jets in the fleet of a privately funded enterprise, the administration and operation of which may be partially funded by and negotiated with government agencies, themselves funded by some means of public taxation.

The operation of the airplane also requires runways, signage and land allotted for these, as well as real estate for fleet parking, maintenance, and administrative operations. Anyone who has been in an airplane knows that weather and air traffic are likely to impact travel scheduled. Jets impact noise levels, influencing health and quality of life issues. Not as obviously, the airplane also affects ground water quality through its exhaust, tire wear, and any fluids that may leak or drip. These "outputs" contribute to (or detract from) the health of the ecosystems throughout its route, too.

Behavior and Function

Returning to the qualities of a system, the purpose of any system is influenced by its function and its behavior. Its structure gives rise to that behavior. The structure of a jet airplane, with its large capacity, is aligned with its purpose of transporting hundreds of people, but remove the turbine or spark plugs and you're left with a large storage locker or perhaps a classroom or greenhouse. With its traditionally streamlined silhouette, extensive control panel, and cramped economy cabin, it's clear that aviation engineers designed it for the purposes of speed, safety, and efficient use of space.

When examining human systems, with their biological needs, psychological impulses, and cultural values, new layers of complexity emerge.

The purpose of a firm's accounting department is primarily to record and review the organization's financial condition, in this way serving the organization's mission to continue to function by having more than sufficient revenues to cover expenses. The department's sub-units for payables and receivables, payroll, inventory, and other assets and liability, are designed with the overall department's purposes in mind. Poor math skills, inadequate knowledge of accounting practices, a desire to present the company in the best possible light in spite of evidence to the contrary, and self-enrichment at the firm's expense are all behaviors that hijack the department's purpose (see more about goal mismatch later in this chapter).

Self-Regulation, Feedback, and Delays

The illustrative airplane is equipped with a set of self-regulatory subsystems. The carburetor regulates the amount of air to mix with the fuel vapor to power the engine. Engine lubrication systems prevent seizing up. The airplane is equipped with feedback mechanisms, too: the instrument panel's lights alert the pilots to take action when necessary, the altimeter helps the pilot adjust the vehicle's altitude, and the radio permits communication about takeoff, route, weather, landing, and other necessary information. But delays are inherent in systems; in this case, an indicator light illuminates before the pilot sees it, and the thrust control speeds up or slows the vehicle down only gradually.

The larger the system, the longer the delays. If more passengers want to take a particular route or flight than there is capacity for them, it may take days for the feedback to get from the ticketing department to the airline's customer service, and weeks to get through to the scheduling system to adjust the airplane size, schedule, or route. If feedback requires a review of policy and procedures or a re-evaluation of assets or services, the delay for changes to expand capacity for new passengers may be even longer.

Self-regulation in organizations is often beset with feedback and delay problems. The small start-up firm is quick and agile, like a speedboat, but the multinational conglomerate resembles more of a huge cruise ship in its ability to process feedback and adjust course. Critical information may get trapped in silos due to departmental barriers, competition, favoritism, or even incompetence. Decisions get bogged down in intra-organizational politics, tabled for further research, or delayed or dismissed due to other priorities. Action on decisions may have to go through committee rounds, wait for assignment or scheduling, or may fall through the cracks entirely.

The Systems Paradigm: A New Point of View

Mindset, or how a thing or a person is viewed, determines how it is valued and treated. In a sense, mindset influences perspective, and perspective creates a version of reality. As mindset itself influences perception, it therefore influences decision making. For example, an investor might be delighted that a recent stock investment has risen 50% in value in a matter of a month or two, until discovery that another stock that had been considered has increased by 200% in the same period of time. Conversely, employees might be upset that salaries have been frozen and benefits curtailed in their company, yet their sense of fortune might change when valued colleagues are laid off. Mindset is inherently relative, a mental model that is built on particular assumptions. Such frameworks are necessary to understand how the world works, and to solve problems. They help in determining what is true and what is false, how to act, what to pursue, and what to avoid.

In the developed nations, the current dominant mindset reduces a phenomenon to a study of isolated parts, is usually explained by a causality that is overly simplified, and assumes that the sum is equal to the parts. Stemming from confidence in mechanics, technology, and hard sciences, many business people rely largely on a mindset that utilizes mechanics, which abridges the relationship between cause and effect.

Here is an example: bacteria reproduce until they cause an infection, thereby causing illness, so killing the bacteria through antibiotics will alleviate the illness. Recently, research has shown that some bacteria, as well as viruses, are essential for health.[23] The emergence and spread of the use of antibacterial products and routine antibiotics applications for non-bacterial illness and in livestock have resulted in deadly antibiotic-resistant bacterial strains. The World Health Organization (WHO) now recommends reform in the use of antibiotics due to increased resistance caused by their prior overuse.[24] Benevolent bacteria are being recognized for their value in medical treatments, as well.[25] Reliance on an overly simple explanation about the parts, rather than the processes and relationships, skewed the perception of value.

Not all phenomena can, or even should, be quantified and reduced, as this denies or ignores substantial spheres of experience and phenomena as well as systemic wholes and interconnections. It also leads to bad theory and policy: "Pretending that something doesn't exist if it's hard to quantify leads to faulty models...No one can define or measure justice, democracy, security, freedom, truth, or love."[26] Spirituality and belief in a universal entity and a unity of all things, whether in connection to nature or not, is not empirically founded, yet many people experience it strongly. Should it be discounted as a misguided human cognitive foible, or allowed for the validity of an as yet unproven or, more likely, unprovable entity? Viewpoints regarding technology, economics, policy, and more, in relation to sustainability, depend on the answers to such questions.

Technology has great potential for addressing environmental, social, and economic issues. Yet, as discussed earlier, technocentrism, the belief that all problems can be solved by technological breakthroughs, is risky. Although humans are a highly inventive species and it is possible that innovations in energy production, emission controls, agriculture, building materials, and water use may alleviate some burdens, the urgency of the situation requires caution.

Our current dominant perspective certainly has value in a number of ways. It has led to an understanding of how parts work, and to theories that are grounded in proof, as well as untold applications used in daily life that most people are barely, if at all, aware of. Yet, it is being demonstrated to be insufficient in adequately explaining or resolving complexity, in science, in nature, in business, and particularly in relation to human behavior.

A systems perspective might be thought of as an upgrade to this way of thinking. It is able to describe observable facts more fully and satisfactorily. For example, it recognizes that everything that is studied is both a system, potentially containing other systems within it, and is a part of a larger, overarching system. A systems paradigm also posits the emergence of capacities and properties within systems and subsystems as they evolve. Broad examples are the emergence of life from the proverbial primordial soup, the development of written language among humans, or the rise of the transnational corporation from ancient bartering practices.

When addressing a problem or question, rather than isolate various parts to identify the source of a specific symptom or occurrence of the problem, the systems mental model examines the system as a whole to identify relationships that might be responsible for generating the observed conditions. Patterns, rather than symptoms, are sought out. If a firm's sales are decreasing, the usual approach might prompt a look at the sales department to troubleshoot the problem. Looking at the bigger picture of the organization, though, there are many possible relationships that influence sales: issues with inventory, distribution, marketing, slow response to market changes, and insufficient interaction between purchasing and production, or something common to all of these.

For example, one behavioral pattern among these individual fault lines points to deficient or withheld communication. The systems thinker does not look to place blame, but to understand how the structure, in this case the information flow within the organization, and between the organization and its market, industry, and society might be creating or contributing to the problem.

This points to an enhancement that a systems perspective brings to thinking about boundaries. Instead of assuming a closed system, one that is isolated, with distinct separation between parts, it assumes an open one. As a result, a systemic paradigm looks at **relationships** within the system, the relationship of the system with other systems in a larger entity, and the relationship of the entity itself to the overarching whole, rather than just at the qualities of the parts themselves.

The current dominant mental model informs us about the qualities of subsystems. Rather than replacing it, systems thinking includes it and enhances it. The added lens of a systemic approach gives a more thorough explanation the world, whether natural, human, conceptual, scientific, or even engineered. Studying the interconnectedness between these topics is one way of recognizing those relationships and an avenue into the systems mindset.

All systems seek some kind of goal, and their parts contribute to the pursuit of that goal. In any living organism the primary goal is survival, with secondary goals of comfort, reproduction, and, in more complex life-forms, resource acquisition. Much the same could be said of most business organizations! In humans and other social animals, this can be understood as status, or position in a hierarchy.[27, 28]

For instance, biologically, our circulatory, pulmonary, digestive, muscular, and other systems all support our bodies' continual survival, just as the human resources, production, accounts, and purchasing departments are designed to do the same for an organization. Actions to reduce fat and increase fitness and attractiveness are comparable to a manager who seeks to root out inefficiencies, reduce expenses, boost productivity, and grab the attention of his superiors by doing more with less. Just as a too stringent diet and overly rigorous exercise program can lead to problems, when goals are pursued without staying attuned to the subsystem's function in serving its super-system's needs, problems develop.

Goal Mismatch: How Assumptions of a Paradigm Guide Behavior

In a business organization, goals of survival and further prosperity guide decision making. There are a number of pathways toward this goal, but the common approach, one with some legal basis, is to improve shareholder value, meaning monetary gain. Related common approaches targeted at improving shareholder value include increasing sales and the value of assets, and decreasing costs, liabilities, and risks.

When these secondary or tertiary goals are unexamined in relation to the primary goals, problems arise. The pressure to increase sales sometimes results in unstable finances because of a need to increase parts inventory coupled with a lag in receivables, decreased organizational productivity due to overworked, unengaged, employees, or an insufficient distribution network. Additional results include bad press because of inaccurate or exaggerated marketing claims and lawsuits arising from the suppression of information that could be detrimental to users/consumers of products or services.

A drive to reduce costs may result in staffing levels or morale that lower productivity, lower-cost parts that break or malfunction more easily resulting in reduced sales or increased returns. Such problems generate internal organizational costs; but these are not the only costs incurred. These are systems issues because the perceived problem that results in the cost reduction goal is very rarely considered within the broader system context of the organization, to say nothing of the industry, the economy, the society, and the planetary systems. Problems construed as stemming from the subsystem generate goals that cause unforeseen consequences, not only within the subsystem, but also in related subsystems and the super-system.

Too often, sustainability practices have solely emphasized short-term and direct economic utility, for which a value is computed in monetary terms. It is only recently that a growing number of economists are realizing there are costs not traditionally considered or included. In Triple bottom line (TBL) accounting practices, externalized costs are integrated for impacts from which an organization does not experience a direct expense, but costs are realized in some other system or subsystem.[29] There are costs of clear-cutting forests, for example, such as the loss of biodiversity, soil erosion and the resultant flooding of downstream land and residences, reduction in the purity of water sources downstream of clear-cut lands, and the loss of wilderness. Yet, the organization doing the clear-cutting does not hold those costs in their balance sheet. Mineral mining done in the developing world, while it provides income for workers, often entails large costs in the localities from which they are extracted, such as environmental damage, pollution, worker health, and disruption of family structure.[30] Again, these costs are usually not accounted for or addressed by the company profiting from the minerals.

Some of these externalized costs are quantifiable, but many costs are cumulative and more regional or global in nature. There is not yet an accepted method to calculate the value of clean air, potable water, sufficient soil, or adequately alkaline oceans. Typically, assets are valued by their market price, but natural assets such as these are not owned or marketed, and there is thereby no framework or formula for calculating values and costs. Alternatively, there are no accepted calculations to compute the costs of widespread and growing poverty, an increasing wealth gap, and the predicted effects of climate change. (Chapter 3 goes into more depth on the economic foundations, social aspects, and environmental qualities of sustainability, as well as an exploration of wealth.)

It could be argued that such costs are not within the control or responsibility of business firms, that they should rightly be externalized and addressed by governments, but this is likely to result only in increased regulation or taxation, which most businesses reject. (Chapter 5 will address why industry is crucial to sustainability.) There are solid reasons, entirely aside from any moral motivation, to account for externalized costs. Recent events, such as the Arab Spring in the Middle East and the Occupy Movement in the United States and abroad, create reverberating political and economic waves that impact companies.

Firms are being targeted either directly by activists, media, and the government, or indirectly through market and social conditions. A social license, the tacit approval of the public, government, and the media to conduct business, is necessary for a firm or industry to exist. The criteria for that social license are becoming more stringent. Regulation is essentially a form of restricting an industry's or commercial interest's social license to operate. Calls for greater transparency and accountability are now more widespread. Insurance companies have begun to require that policyholders have plans to address environmental and social risks, with institutional investors not far behind. Viewed from the narrower lens of shareholder value alone, corporate sustainability efforts are no longer optional.

Examining issues through a short-term time frame is characteristic of the current dominant mindset. Financial success is judged on quarterly performance and annual gains. Exponential growth is rewarded but, as will be discussed in the next chapter, emphasis on rapid and explosive growth in the near future frequently has had a negative impact when viewed over an extended span of time. If businesses wish to remain competitive and serve their shareholders, they will be required to contribute to social well-being in the long-term.

Chapter Summary: Key Takeaways

The business world, as well as society, has reached a tipping point: gone are the days when security and the promise of prosperity were taken for granted. A host of economic, social, and environmental incidents have forced industry to begin to address how they conduct business, and the impacts of their practices and policies on society and the planet. The evidence of this trend is the tremendous growth in sustainability practices and reporting by businesses of all sizes.

Research on the cause of the problems now being faced reveals the need to develop an updated worldview, one that is more adequate in explaining the reason behind those problems. Studies have shown that this revised paradigm, based on a principle of interdependency between systems, has gained traction, not only in science, but also in many disciplines, and in commerce.

In the broader view, sustainability is the ability for economic, social, and natural environmental systems to coexist in a dynamic equilibrium far into the future, the capacity to meet current needs while ensuring the ability of future generations to meet their needs. Technical solutions developed using an outmoded perspective will necessarily fail, as they will continue to result in the same kind of issues that currently exist.

A system is a self-organizing entity that generates inflows and outflows. Within the system are hierarchical, differentiated subsystems which act in interdependent relationship. Central control is balanced with subsystem control, providing resilience. Subsystems that are misaligned with the purposes of the overarching system create imbalances, resulting in destabilizing feedback loops, mechanisms by which the system determines a need to change behavior in order to maintain a dynamic equilibrium. Increasing size creates increasing delay in observable change.

Each of these systemic qualities is present in any business and industry, as they are in all biological systems. Just as business has evolved from highly local tribal bartering to immense and widely distributed transnational commerce, human thinking is in the process of evolving from a mechanistic and reductive worldview to one informed by these principles. The ability to think in terms of patterns and relationships, rather than simply causes and parts in isolation, is becoming a critical business skill.

CHAPTER 2

What Is Meant By Sustainability and Who Defines It?

The previous chapter briefly introduced the concept of sustainability and the paradigm of systems thinking. Distinguishing the systems mindset from other mental frameworks, Chapter 1 offered it as an enhancement to the current mindset that has been predominant and unquestioned until relatively recently. A variety of examples were used to explain what is meant by a systems paradigm and illustrated how such thinking is more comprehensive and sufficient in explaining and addressing the challenges of this complex world, and in business concerns more specifically. This chapter expands on the concept of sustainability and how it bears on business practices, incorporating the systemic mindset. Triple bottom line (TBL), CSR, stakeholders, and other often used terms are explained within the context of systems and sustainability.

Sustainability Jargon 101: TBL, 3P, EEE

It wasn't until 1987 that the word sustainability, as used in business, meant anything more than the continued survival of a commercial enterprise, operating with sufficient financial health to avoid bankruptcy or outright demise. That year the United Nations World Commission on Environmental and Development issued a report that would become the foundation for the ongoing discussion about sustainability. Known as *The Brundtland Report*, it addressed, and defined, sustainable development as "the ability to...ensure that it meets the needs of the present without compromising the ability of future generations to meet their own needs."[1]

Although the report was extensive in explaining the causes of environmental degradation, particularly poverty and increasing social inequity, the above-quoted sentence about intergenerational responsibility became the baseline for the concept of sustainability. Unfortunately, like most quotes intended to simplify a term, it was taken out of context, leaving out many of the crucial components of sustainability (and sustainable development) discussed in the report.

While focusing on parity between generations, this definition does not address equity *within* generations. Three pages earlier, the report cites the gap between developed nations and developing nations, with the former effectively making the rules and using the majority of the ecological capital, and states, "This inequality is the planet's main 'environmental' problem; it is also its main 'development' problem."[2] Reinforcing this contention, it later states, "It could be argued that the distribution of power and influence within society lies at the heart of most environmental and development challenges."[3] The report was written by scientists and development experts. Its purpose was to address the current and future concerns of all humanity. It pointed out that a minority of humanity lives in industrialized nations, and the comparative percentage of that minority was dropping.

This social equity aspect is one expression of the systemic nature of the concept of sustainability. In a broader sense, sustainability indicates the interrelationships between pretty much everything: economics, industry, people, resources, ecosystems, culture, policy, government, science, technology, management, ethics, and more.

The Brundtland Report's analysis notes six sustainability concerns: population and human resources, food security, species and ecosystems, energy, industry, and urban challenges. Human population growth is forecast to continue through the middle of the 21st century. Using population in relation to food security offers a starting point to illustrate the systemic nature of sustainability. Over 900 million people around the world do not have sufficient nourishment, and this problem is projected to grow as population increases.[4]

A host of factors influence the ability to address this set of current and predicted future problems. Agricultural output is a technological issue, but is also increasingly recognized as an ethical and biological one if genetically modified seeds, plants, or animals are used. It is a cultural (and often technological) issue if local knowledge and skills are not respected and utilized. It is a social equity issue if segments of populations are pushed to marginally productive land by government or agricultural and development industry lobbying influence on government.

Agricultural output is also an environmental issue if crop production results in soil erosion. Water and soil pollution are caused by chemical runoff from fertilization and pesticide and herbicide applications. The production of methane in animal husbandry contributes to climate change. An additional systems impact is the loss of species biodiversity, a rich array of potential materials and solutions to existing problems, through the clearing of sensitive habitat for agricultural purposes results.

Beyond this, it is also a policy and economic issue in terms of agricultural subsidies for either locally produced or imported foodstuffs, land use reforms, distribution logistics, and other types of government policy and support for farmers and the agricultural and food production industry. To some extent it is also an energy issue, and an industrial one, through the production, transport, and use of equipment, fertilizer, and other chemical applications in the growing of crops and livestock, and the processing, packaging, transport, and protection of foods from spoilage.

Climate change provides another systemic illustration of a seemingly environmental problem that is broadly embedded in many human systems and processes. Although there are some naysayers,[5] the consensus is that the phenomenon is human-generated.[6,7] It is caused largely by commercial and personal transportation and industrial and domestic energy use, as well as some agricultural practices, and so has its origins in energy production and use and industrial agricultural processes.[8,9] The cutting down of forests for agricultural production and cattle grazing reduces the available tree mass that absorbs carbon dioxide, one of the gases responsible for warming.

Resolving these problems, then, requires examining their systemic impacts from a multidisciplinary approach, as the problems are generated from embedded issues in intertwined subsystems of the larger biosphere.

A philosophical or moral inclination toward social equity is actually beside the point. Even as the middle class in developing nations such as China and India rapidly expands, a growing percentage of the world's population is becoming increasingly marginalized. The wealth gap has been widening between the least developed and other countries, and is forecast to continue to grow.[10] The ensuing social unrest and political upheaval make global trade more difficult and risky.

Developing nations are not alone in experiencing such pressures, and the Occupy Movement may be an indication of this trend. In 2011, the Organization for Economic Co-operation and Development (OECD) reported that income inequality has been rising in OECD countries (industrialized nations) since the late 1970s. That gap continues to widen in most of those countries, with the income of the wealthiest 10% increasing more rapidly than the poorest 10%.[11] The pace is widening more rapidly in the United States than in most other countries, caused by factors including technology, globalization, and government inattention.[12]

Is Sustainability an Outdated Concept?

As a term to describe the continuous march of civilization while avoiding catastrophic environmental destabilization, massive social upheaval, or severe economic devastation, and at the same time also addressing the life-threatening unmet needs of at least one seventh of humankind, sustainability has been serving for several decades. As thought and research on the subject has progressed, there are several streams of criticism to be aware of.

Among the louder set of critical voices are those who believe that the term has been co-opted by marketers and firms that skim off the benefits of its use while ignoring or brushing aside anything but its most superficial applications. Sustainability is seen as a fashionable idea of the moment rather than as a complex set of interconnections with lasting and deep value.

These critics have a salient point. Not to denigrate small incremental changes or those that immediately and clearly benefit an organization, but the term has certainly been used as a ploy to gain advantages while firms are simply conducting business as usual, or worse, deceptive tactics to look responsible or claim to adhere to certain standards while doing otherwise. Even well-intentioned firms make commitments but often fail to follow through, according to recent research.[13] This is also a complaint about sustainability reporting: to report without setting goals, to fail to report on progress toward previously set goals, to simply gather data without creating an action plan for change, or to check off indicators without transparency or accountability is to engage in a form of greenwashing.

The original term from the Brundtland Report, "sustainable development" has been faulted as an oxymoron, and as a misguided attempt to turn the developing world into a copy of the consumer model economy of the developed nations. Indeed, the "sustainable consumer" shares this contradiction of extending the problem rather than correcting it.

Another set of objections, which apply to both the concept of sustainability and the term, faults the word for being past its usefulness, partly for the reasons above, but also because it is overused, too vague and elusive, doesn't inspire, and evokes a sense of fatigue. Newer terms meant to prompt a more enlivening and paradigm-shifting sensibility include "flourishing,"[14] "thriving,"[15] and "abundance."[16] Even while sustainability may be a concept that has arisen as a result of a new skepticism toward the notion of progress and as a dawning question as to whether the world is one of scarcity or abundance,[17] the backlash centering on the term is not about whether or not there is an urgent set of issues to address but rather how to address them.

The Three-Legged Stool

For the purposes of business interests, though, the concept of sustainability has often, and thankfully, been distilled into a simple three-legged stool. The TBL or 3BL, an accounting framework designed to integrate ecological and social performance into financial reporting, is a means for companies to incorporate consideration for people, planet, and profit, or 3P. Alternatively expressed as 3E, for equity, ecology, and economy, it's a notion intended to include natural and human capital to measure organizational success. It is not a fully accepted path to value creation, but the premises underlying TBL (and the other concepts that follow in this section) are increasingly used not only in business settings but also in governments, regional development initiatives, and non-profits. While Andrew Savitz and Karl Weber brought the concept into the business market by associating it with improved profitability,[18] John Elkington, author of the book *Cannibals with Forks* (1997), popularized the concept, insisting that the drive toward sustainability provides an unprecedented opportunity for commercial gain and competitive advantage.

Elkington explores what he calls "paradigm shift revolutions" between the existing capitalist mindset and that informed by TBL. Values will move from a preeminently hard, quantitative, sharply economic standpoint to grow more inclusive, where softer qualitative values such as stakeholder relationships, cultural diversity, quality of life, and a stronger consideration for the future are incorporated. Transparency, which has traditionally been closed, will become more open in what he called the "X-ray environment" as businesses and government find that keeping secrets becomes both more difficult and more costly, with whistleblowers and WikiLeaks-type organizations becoming more common. Lifecycle technology is being recalibrated from a focus on only the product to the entire cradle-to-cradle value chain, where production processes are identified and examined for the impacts they cause in heretofore seemingly unrelated outcomes (tobacco and lung cancer, hydraulic fracking and ground-water contamination, carbon dioxide and climate change).

Partnerships between organizations, whether intra- or inter-industry, or between business and government and non-governmental organizations will transition from subversion, a highly competitive state, to a more cooperative, symbiotic one, which Elkington labels "co-opetition," in which loyalty and trust are built and maintained. Time horizons will switch from a wide view of current circumstances as informing knowledge to a longer perspective that uses the past as a lesson and the state of the future, generations ahead, as one of the most important considerations in current decision making. Where the old paradigm of governance has been exclusive, a shift toward the inclusion of multiple stakeholders and greater diversity of backgrounds on boards is another of the revolutions in mindset he sees as both desirable and inevitable.

One of the issues cited as a problem with TBL accounting is that the three components do not share a metric. Profit is measured in monetary terms, but there are not yet valid ways to quantify and compare the loss of critical ecosystems, clean air, or healthy communities. Another problem is adjusting for industry sector operational differences. Even if an index were to be created, there is the problem of how much weight to give to each component.

Yet, these issues do not negate the value of TBL. While there are currently no universal means of measuring TBL, each of these criticisms can be addressed by businesses through stakeholder engagement (see the section on stakeholders later in this chapter). Participation from stakeholders and subject matter experts can help a business determine the set of measures to be used. For example, environmental impacts that are quantifiable include consumption of resources, such as energy and water, and pollutants emitted. Quantifiable social impacts include worker safety, diversity, employee training and development, fair dealing, anti-corruption, and community investment.

A growing industry of sustainability consulting firms assists businesses in creating systems for measuring, tracking, and reporting these metrics for businesses at the start of the learning curve. Reporting may be accomplished in a variety of ways. Some firms incorporate TBL information into their annual financial report, others issue an environmental or social report (or both combined); still others issue a full sustainability report.

Prior to deciding about whether and how to begin TBL measurement and reporting, planning is strongly recommended (see Chapter 7 for more detailed information). Planning and some research are necessary as there may be national and international reporting trends and standards within your industry. Any kind of sustainability planning and reporting will require resources and budgetary considerations, and board and C-level support, which is usually gained by establishing the case and setting high level objectives aligned with the organization's values and mission. A reporting strategy and direction will need to be established, and the engagement of stakeholders will help the firm prioritize concerns and identify appropriate performance indicators. As always, there are legal implications to review. Gaps and barriers between current and desired reporting should be determined, as well as the content and structure of the report, specific data to be collected, and methods for data collection and analysis. The Group of 100, an association of senior finance executives in Australian firms, offers a useful guide to TBL reporting.[19]

The Global Reporting Initiative (GRI), an organization that has developed a sustainability reporting framework, is among the more commonly used methods of documenting TBL impacts, which will be discussed in Chapters 6 and 7.

Another aspect of reporting is verification by a third-party auditor to assure the report's credibility and reliability. Some metrics and reporting standards require third-party verification while for others, like GRI, it is voluntary.

A host of terms and acronyms have evolved to capture corporate interpretations for instituting efforts and reporting on results, although there are differences among them. (Refer to the glossary at the beginning of this book for a list of terms and acronyms.)

CSR and Sustainability

Corporate Social Responsibility (CSR) is among the most popular of terms to describe how businesses address ESG issues. There are some federal, state, and local laws regulating how industry conducts itself in regard to impacts like air and water pollution, fisheries management, and pesticide use, employment policies, working conditions, and other aspects of commercial conduct. Yet, there are no regulations regarding what is or is not included in CSR activities or reporting.

It is, at least at this time, a legally voluntary set of practices, and firms are at liberty to employ the term however they choose. One definition of CSR is "the way in which business consistently creates shared value in society through economic development, good governance, stakeholder responsiveness, and environmental improvement."[20]

As a result practices that are labeled CSR run the spectrum from marginal to groundbreaking. Some firms believe that they are "doing" CSR by such minimal initiatives as allowing employees to take a day off to volunteer or funneling some charitable donations into their local community. Others restructure their firm's mission, vision, and processes to incorporate social, environmental awareness, or both, and deliverables into all aspects of their operations.

However, it must be understood that CSR efforts are not the same as sustainability. For one thing, sustainability is a hoped-for state rather than a current fact. While scientific research tells us that current resource use is unsustainable, it is less clear as to exactly what it will take to reach a sustainable state of dynamic equilibrium. Aiming toward net zero resource use, which means bringing the overall consumption of resources down to an effective rate of zero, is one way to address the environmental side of sustainability.

Social sustainability, the notion that all individuals in current and future generations should have access to social resources that allow them to develop their capabilities, encompasses human rights, labor rights, and corporate governance. From a policy standpoint, the aims of social sustainability include basic infrastructure and services, education, employment, freedom, justice, equitable distribution of power and resources, and access to decision making. The extent of government involvement in the provision of these ends is a contentious issue, but industry's contribution is even more controversial. Even so, companies can and do make efforts to minimize harm to their employees, the communities and supply chains within which they operate, and consumers of their products or services.

Secondly, and perhaps more importantly, sustainability requires a focus on all three legs of its stool without prioritizing one or two over the other(s) to the point at which, figuratively, the stool is unstable. The metaphor portrays the need for balance between economics, environment, and social equity. As global trade has emphasized economic growth above all other considerations, this imbalance has manifested itself in costs to the global biophysical resource base and social well-being to the extent that one or more of the subsystems may well deteriorate beyond the capacity to recover. As much as people may muck it up, the planet will eventually regain its dynamic equilibrium. The "anthropogenic" (man-made) influence, however, may create conditions that make survival for humans and much of other life, including the plants and animals that society relies on for nutrition and stable water resources, much more difficult than it is currently.

Environmental and social doomsaying is both annoying and depressing. Civilization cannot afford, though, to ignore ever accumulating research indicating deterioration to air, oceans, biodiversity, soil levels, potable water supplies, climate stability, political equilibrium, and the security threats each of these provokes. As Ray Anderson, the now deceased CEO of sustainability innovator Interface, asked, "What is the business case for ending life on earth?"[21] In addition to the media and the public, insurance companies and investors have begun to believe the answer to this question is a firm "There isn't one!"

CR, CSR, Green, Sustainable—To What Degree?

Some large public firms saw this impending trend and became leaders and innovators at implementing practices.[22] Regularly appearing since 2005 on the annual Global 100, a list of large cap firms with the strongest performance using 12 key performance indicators, is Intel. Their multi-pronged approach includes intra- and inter-organizational strategies, as well as creating industry level initiatives and partnering with governments. One strategy that improves buy-in across the organization is tying the variable compensation portion of all employees' pay to the achievement of their sustainability metrics. Intel is working with a number of organizations in the computer industry to develop and market more energy efficient computing devices through the Climate Savers Computing Initiative, and on better social and environmental extractive practices through the Electronics Industry Citizen Coalition. The corporation also worked with governments to develop environmental and social impact standards.

Simply being listed does not, of course, guarantee that a firm is free of risks. Some of the corporations appearing on these lists have simultaneously been involved in fraud or corruption scandals. Some have engaged in highly public dubious environmental and social practices. Sustainability performer listings are subjectively designed and can be susceptible to influence. Some firms may pursue these listings as a tactic to allay public scrutiny and outrage.

CSR is not the only way organizations characterize their efforts. Corporate responsibility (CR), corporate citizenship, green practices, and corporate accountability are other frequently used terms. Accountability and transparency are relatively new to the business world, so the use of terms is still developing and is in a state of flux. Some companies interpret corporate citizenship as solely addressing governance issues. Other firms consider charitable giving as the appropriate totality of their CR efforts. Green practices are likely to refer to a focus on environmental efforts, with little attention paid to the social and governance components of sustainability. At this stage, these terms are used somewhat interchangeably amongst firms for any programs beyond traditional short-term economic concerns. While individual firms might have more specific and limited definitions of them, this book defines sustainability, in the business context, as encompassing a comprehensive set of governance, economic, social, and environmental factors.

Sustainability and Stakeholders

Corporate performance has typically predominantly been measured by financial return to shareholders, as corporations have a legal responsibility to increase shareholder value. In the last quarter of the 20th century, an influential article in *The Journal of Financial Economics* pushed this motivation to the forefront of corporate attention.[23] According to the authors of the article, because executive pay was not tied to shareholder value, the executives had more incentive to attend to their own pay instead. Consequently, not only was shareholder value diminished, the entire economy was not getting adequate value from a stronger focus on increasing company value. This realization was soon followed by a practice of tying executive pay to stock value, in the form of stock as compensation.

Unfortunately, this scheme was designed in a way that encompasses only short-term moves in stock rather than longer-term shareholder value. Executives do not have as much control over long-range results as they do over the next couple of quarters. As Roger Martin cites in a 2011 *Harvard Business Review* article, the data indicates that three decades of maximizing shareholder value resulted in lower annual real returns than were experienced in the period 40 years earlier.[24]

The positing of shareholder maximization as the desired end result of corporate function serves as a useful application of systems concepts. **Remember that the purpose of a system determines its behavior.** Most corporate mission statements don't prioritize, or even mention, maximizing shareholder value, yet most firms operate as if their purpose is exactly that. Mission statements, often also called "statements of purpose," have firms dedicating their efforts to being the best in their industry, excelling at customer satisfaction, or creating and inspiring progress. When the purpose-in-use is to maximize shareholder value, product quality and customer service may suffer, although operations based on mission statement purpose do not necessarily mean that profits and share price will decline.

That is only the first systems lesson here, though. Shareholders and customers are only two groups among many upon whom an organization's purpose will have an impact. Just ask the employees of General Electric (GE) between 1981 and 2001 when Jack Welch was the CEO, or Gulf coast residents and businesses after the BP Deepwater Horizon oil spill in 2010.

Over the last few decades a steady succession of corporate disasters, such as Love Canal, the Union Carbide disaster in Bhopal, the Exxon Valdez oil spill, the WorldCom and Enron accounting scandals, and Nike's sweatshop practices, have highlighted that businesses do not exist in a vacuum. These incidents drew significant media attention, followed by public outrage, and sometimes, by legislative action. Calls for greater accountability and transparency to non-shareholder stakeholders became widespread.

Shareholders and Stakeholders

The term stakeholder originated at Stanford Research Institute in the 1960s, but was taken up and popularized in R. Edward Freeman's 1984 book *Strategic Management: A Stakeholder Approach*. Defined as individuals and groups who either benefit or are harmed, or whose rights are respected or violated by the firm, Freeman included stockholders, employees, managers, customers, suppliers, and the local community. His question, "For whose benefit and at whose expense should the firm be managed?" implicitly referenced a systemic view of the corporation. Value creation is still the purpose of the business, yet that value should not be created at the expense of a subsystem.[25]

Later discussion about stakeholders by some management experts expanded the stakeholder groups to include future generations and the natural environment, thereby widening not only the breadth of the system but also its time frame. The consensus of scientists has concluded that, as a species, humans are benefitting at the expense of both of these stakeholders. There is little point in organizational longevity if the survival of future generations, who make up employees, customers, and vendors, was imperiled because the current living generations did not take decisive and effective action.

Even if you do not agree with a moral obligation toward future generations, including your own descendants, consider your dependency upon natural capital for basic survival. The era of abundant potable water, ample arable soil, and taken-for-granted clean air is behind us.[26]

Beyond existing generations, there are reasons to take decisive and substantive action, though. Climate scientists suggest that even if greenhouse gas (GHG) emissions were to not increase beyond 2007 levels, the earth will continue to warm.[27] Studies project that climate change is likely to lead to more extreme temperatures at both ends of the spectrum, disruptions in the food supply, drought, and higher incidence of infectious disease by the end of this century.

Critics of stakeholder theory assert that it is impossible to maximize value to more than one stakeholder group simultaneously, and that value creation is more of a scorecard than a purpose. Indeed, a number of scholars who support the general idea of stakeholders believe the theory inadequate in explaining how to weigh and prioritize stakeholder interests. In spite of these suggested flaws, many companies have engaged in stakeholder management, balancing shareholder value maximization with value creation for other stakeholders.

Stakeholder Inclusion Examples

The term "stakeholder inclusion" is often seen in the context of governmental agency and intergovernmental development projects, as a strategy for improving the success of an initiative, but can just as effectively be applied in a business context. Engaging stakeholders by informing them through communications channels and by soliciting their feedback about processes, programs, products, and plans is a systemic business strategy. It serves as an early warning mechanism to avoid problems and concerns, and it increases stakeholder buy-in while polishing a firm's reputation, enhancing brand loyalty, and bringing diverse perspectives together for innovation and creativity.

Like it or not, stakeholder activism is driving corporate change. In April 2012, Coca-Cola withdrew its funding from the American Legislative Exchange Council (ALEC), on the heels of social-media-generated negative publicity about ALEC's promotion of voter ID legislation.

Critics of ALEC, including stakeholder groups such as customers, NGOs, and communities, asserted that this legislation was a way of suppressing votes, particularly among minority populations, that ALEC promoted other legislation that would disenfranchise minorities, and that they promoted climate change denial legislation.[28]

Firms are finding higher levels of consumer distrust as well as higher expectations. Recent research found that only 44% of Americans trust companies' green claims, and that 77% would be willing to boycott a firm if they believe the firm had misled them. Close to 90% expect firms to understand and address the environmental lifecycle impact of their products, including manufacture, use, and disposal, and they want more easily accessible information for making decisions.[29] A study of executives and managers in 113 countries indicated that greater levels of stakeholder inclusion, even including collaboration with competitors, were practiced by firms that were profiting from their sustainability practices than those firms that were not.[30]

Coca-Cola's stakeholder inclusion efforts go beyond the usual customer complaint line and employee feedback surveys. They proactively and regularly meet with trade and labor unions, host annual industry conferences on human rights, convene expert panels to address child labor issues, and have been active in forming an industry association to promote responsible supply chain sourcing.[31]

As one of Starbucks' sustainability efforts, a program to reduce post-consumption coffee cup waste, the firm has had annual "Cup Summit" events. Initial actions to address this waste by creating recyclable cups met with challenges due to variances in local recycling capabilities where the stores were located. The step Starbucks took to resolve the problem was to invite a wide range of stakeholders, including recyclers, municipalities, cup manufacturers, raw material suppliers, NGOs, academics, and retail and beverage businesses to come together to develop solutions.[32]

Why Sustainability Is More Than a Trendy Concept

A Massachusetts Institute of Technology (MIT) Sloan Management research study released in December 2011 concluded, "Sustainability Nears a Tipping Point."[33] Two-thirds of the nearly 2900 managers and executives who responded to their survey agreed that sustainability was a critically important competitive issue in today's marketplace. While only about a third of responding companies were currently profiting from their sustainability practices, an increasing percentage of firms were investing in such practices. A critical mass is building, and the question is shifting from whether to incorporate sustainability to how to do it.

The study cites both external and internal pressures as driving the trend toward sustainability. Consumer demand, media, social networking, government regulation, higher expectations from institutional investors and insurers, and CSR rankings are some of the external drivers. Firms that profit from their practices also usually are internally motivated through board directives, executive compensation tied to sustainability metrics, employee activism, clear metrics, and organizational support for innovation in not only practices, but also business models.

These drivers have created the push toward sustainability in business, but the MIT Sloan study found convincing evidence that both opportunity and necessity are compelling companies to do more than consider sustainability initiatives as an add-on. Instead, sustainability is progressively driven deeper into all the units, processes, and practices within a firm, and a systems perspective is the common thread that unites them.

A number of business management experts are coming to the conclusion that reporting on ESG practices, in addition to the required public financial disclosure, will soon be mandated. Harvard Business School's Robert Eccles believes that integrated reporting will not only be mandatory, but will also be standardized and enforced.[34]

Integrated reporting means that the firm uses that process as a mechanism, even a discipline, in positioning the ESG aspects to broaden their overall value creation strategy, rather than as an add-on. The adoption of integrated forms of reporting is gaining momentum. The European Union and South Africa are have mandates on the books or pending, and many multinational firms are adopting the practice, in the process pushing their industries, vendors, and suppliers in that direction as well. Sooner or later, Eccles believes, legislation in the United States will follow.

There is also pressure for integrated reporting from investors, who increasingly perceive non-financial information as material to firm longevity and profitability. Governments and NGOs are also more often and consistently pushing companies to be transparent about their ESG practices, according to a 2012 article from Bloomberg Businessweek.[35] Despite the fact that such reporting isn't yet mandatory in the United States, every year more firms have made these disclosures, seeing them as a long-term profitability strategy.

Consumer activism and non-governmental groups are another set of drivers. For example, in August 2012 consumer products giant Johnson & Johnson announced a web-based ingredient transparency initiative and the removal of certain ingredients from their products, citing consumer concerns about potentially toxic ingredients.[36] Although currently considered safe by regulatory agencies, parabens and ingredients that release formaldehyde were of particular focus by the Campaign for Safe Cosmetics, a coalition of NGOs including the Environmental Working Group.[37]

Over the last few years, fast food company McDonalds has bowed to consumer and activist group pressure on a number of practices. They discontinued selling chicken fed on soya from plots of deforested Amazon rainforest, made their Happy Meals healthier, halted the use of genetically engineered potatoes, replaced foam cups with paper ones, required their pork suppliers to stop using gestation crates, and ceased using "pink slime" in their ground beef.[38]

Campbell's is phasing out its use of bisphenol-A in their can linings over consumer fears about the substance.[39] After a consumer firestorm, Bank of America dropped a proposed $5 debit card transaction fee.[40] Starbucks began selling fair trade certified coffee.[41] Public demand shapes decision making in business, and as the public perception of acceptable business practices shifts, so does the corporate incentive to change.

Insurance regulation is in the process of becoming another force impelling firms to address their operations. State insurance commissioners in California, New York, and Washington State now require insurers to switch from a backward-looking risk assessment model to one that anticipates how they will respond to predicted climate change impacts, such as increased wildfires, rising sea levels, and more frequent and severe storms.[42] It is inevitable that these insurers will require their policyholders to address these risks in turn or face substantially higher premiums.

Built-in Sustainability Benefits to Firms

A number of built-in benefits to integrating sustainability programs have already been discussed, yet cost savings, improved image, enhanced employee attraction, retention, and engagement, and new markets and revenue sources are only some of the potential wins. Another bonus is that measurement activities required to collect data for reporting can, and often do, lead to product, service, and operational innovations and improvements and more effective business processes.

Incorporating sustainable practices often strengthens compliance with legislative and policy mandates, and reduces exposure to tax issues. Sector leaders have the opportunity to be influential in the creation of government policy, which then provides them with a competitive advantage. Addressing sustainability risks often leads to improved access to investment capital. Enhanced goodwill can also increase speed to market by reducing opposition.

Seeking solutions to sustainability problems drives innovation, which then usually drives value creation. Sustainability is, in essence, basic business pragmatism. Reducing dependence on foreign oil is not only a national security issue, but reduces operational risks and energy price uncertainty for industry. Ensuring the protection of not only the viability of critical domestic ecosystems and future prosperity for the current generation's children and grandchildren, a focus on sustainability in business practices elevates a nation's economic standing among other nations.

As if these reasons weren't compelling enough, from a business case perspective, there is research documenting the superior economic performance of companies that have integrated ESG or sustainability strategies.[43-45] In one study, firms considered sustainability leaders in six different sectors outperformed the general stock market by 25%, with 72% outperforming their peers.[46] Firm value has been shown to be higher in companies with CSR strategies that also have high consumer awareness.[47] A 2011 study by researchers at Notre Dame and Georgetown found an association between firm value and carbon emission levels, where heavier emitters had lower value. It wasn't a small hit, either: for every additional thousand metric tons of carbon emissions for the sample of S&P 500 firms, firm value decreased by an average of $202,000.[48]

Large cap firms aren't the only beneficiaries, or practitioners, of sustainability practices. Small and medium-sized firms are paying attention, and are also realizing bottom-line benefits. A survey of over 1300 small to medium enterprises from the United States, United Kingdom, and Canada found that the majority of these firms were defining and developing sustainability strategies.[49] The reasons were little different from those of larger firms.

The definition of sustainability and application of its underlying principles to business models, processes, and practices certainly vary from business to business. Those firms recognizing themselves as an interdependent entity, a system with interlocking subsystems operating in a set of wider enveloping systems, are more effective and successful in their sustainability efforts. They are leading the way, setting the standards for transparency, accountability, metrics, and depth of reporting, selecting frameworks and developing best practices that laggards will be expected to follow.

The next chapter delves deeper into these components of sustainability and how they interact in the larger global system. It starts by examining how certain taken-for-granted assumptions lead to erroneous thinking and less-than-optimal outcomes.

Chapter Summary: Key Takeaways

While the concept of and practice toward sustainability are contested, most experts agree that any definition and application require attending to a balance of economic, social, and environmental concerns, in both present and future contexts. This triad of components underlies the foundation of TBL thinking and accounting practices, as well as most sustainability frameworks. Sustainability is a systemic concept with an expanded time frame, incorporating an understanding of how civilization has reached this point and endeavoring to create parity for the current as well as future inhabitants of the planet.

CSR and sustainability efforts are currently voluntary from a legal perspective, but are becoming part of a social expectation. As impacts and issues vary between industries and firms, there is no single best way to "do sustainability." Programs and initiatives run from the substantive, wide-ranging, and deeply integrated to shallow greenwashing. A systemic approach requires awareness of and interaction with stakeholders, those parties impacted by the firm, positively, negatively, or both, such as employees, vendors, NGOs, local communities, government, customers, and distributors.

A number of pressure points are driving responsible business practices, including public sentiment, media exposure, activist campaigns, government regulation, institutional investors, and insurers. The good news is that there is a solid business case supporting those practices: cost savings, managed risks, reputational enhancement, customer loyalty, employee engagement, and new revenue streams and markets are a few of the payoffs.

CHAPTER 3

Systems and Sustainability

Chapter 1 introduced the concept of a systems framework as an enhanced and more effective mental model with which to understand the complexities present in the world and specifically in business today. The notion of sustainability was introduced, and was then further developed and tied into the systems perspective in Chapter 2, providing a foundation for illustrating how these related concepts are creating a shift in the world of global commerce. This chapter delves deeper into the three aspects of sustainability, the economic, social, and environmental, and their impact on commerce. It starts with a look at how conceptions of wealth are related to both systems and business.

Examining Economic Assumptions

Does Growth Benefit Everyone?

The media feature stories about "the economic system," but the understanding of it changes when viewed through a systems lens. The predominant current view of the economy is that it operates by rules that are well understood, and that policy makers and economists manipulate the variables within those rules for everyone's benefit. That view also assumes that the economic system's purpose is to continue to grow at as fast a rate as feasible in order to create wealth for all, summed up in that aphorism, "a rising tide lifts all boats." However, since the income gap has been steadily increasing rather than decreasing, this assumption must be called into question.[1]

Evaluating that assumption more deeply leads to some problems. As management expert Henry Mintzberg and colleagues point out, a rising tide does not rise forever, and boats that are moored will be inundated, as will everything in low lying areas.[2] When a tide has risen above normal levels it becomes a flood; when it cycles out, ruin is left in its wake.

With an indefinitely rising tide—equivalent in this metaphor to unceasing economic growth—only those who have made it to the highest elevations are safe. A sad and ironic parallel is that climate-change-induced rising sea levels may be imperiling the most vulnerable populations residing in coastal areas, amounting to 10% of the world's population—and some predictions include impacts to coastal cities in the United States.[3]

One significant difference between the standard conceptualization and a systems perspective on the economic system is a system/subsystem hierarchy. Business experts, scientists, and even economists are realizing the flaws in positioning the economy as the bigger and more important system within which people and the natural world exist (see Figure 1.1 in Chapter 1). As economic progress has fueled progress in lifespan and quality of life, as discussed in previous chapters, these measures are somewhat lopsided and come at a cost to people and other planetary life. Table 3.1 offers some useful distinctions between current assumptions and systems assumptions of economic components.

The Myth of the Economic Man

Another component of economic theory that begs re-examination is the view of "economic man" as a rational decision maker.[4] The marketing department of any firm is a testament to the fact that consumers do not

Table 3.1. Economic Assumptions

Component	Current assumption	Systems assumption
Breadth of benefit	Unceasing growth benefits everyone	Unceasing growth benefits a small segment
Individual as	Economic (rational) man	Using both emotion and logic
Goal	Limitless growth	Prosperity for society and ecosystems
Resources	Substitutability of monetary and human capital for nature	Losses in nature cannot be made up for by monetary gains
Depth of benefit	Progress measured by single quantitative monetary gain	Progress measured by multiple qualitative gains (employ-ment, literacy, life expec-tancy)

depend exclusively or dependably on rationality in making purchasing decisions. While people use a combination of logic and emotion, the former is clearly not the basis for most choices.

The economic man theory also does not account for values or inclinations other than self-interest. Examples of greed may abound, yet the growing push toward CSR and corporate citizenship, as well as the rise of social entrepreneurship are a few examples of evidence that belie the notion that self-interest is the only guide for every person and in any economic transaction. The theory recognizes the individual, but fails to account for the fact that commerce (and life) is inherently and fundamentally social and built on trust. Competition is a strong driver of all life, and humans are no exception, but cooperation is just as elemental. Without it there would not be families, communities, corporations, or nation states.

The Myth of Limitless Growth and its Sustainability Implications

Assessing the assumption of unbounded economic growth from a resource perspective leads fairly quickly to the realization of its flaws. The planet's resources are finite, and economic growth is not possible without continued supplies of resources. Of course, one business opportunity exists in radically improving resource efficiencies, and this is slowly happening. There are two constraints on eco-efficiencies, though. The first is a growing population, and the second is the rebound effect.

The rebound effect is a response to technological and income improvements that minimize the benefit of resource efficiencies. Homes have become larger, from a median size of 1500 square feet in the 1960s to about 2200 square feet during the period 2004 to 2009 in the United States.[5]

The number of homes with air conditioning has increased from 56% in 1978 to 72% in 1997, and the percentage of homes cooled with central air increased from 23% to 47%. People have been using their air conditioners more often, too. Even as efficiencies in new air conditioners improved in that time by 20%, household electricity usage for air conditioning rose by 35%. Larger homes with bigger rooms cooled by central units meant that more space was cooled, leading to higher energy use.[6]

As income and living standards in developing countries improve, more people are building homes, buying cars and appliances, and the rate of use of resources and energy continues to climb. That doesn't mean resource efficiencies should not be pursued—indeed, they need to be, vigorously and expeditiously. What it does mean is that resource use has continued to grow despite efficiency technologies.

One model that revises economic theory is called Natural Capitalism.[7] In this model, all economic activity is informed by four principles: radical resource productivity, biomimicry, a service and flow economy, and investing in natural capital.

- The radical restructuring of resource use, tripling and quadrupling efficiencies, lowers resource extraction and waste pollution, and has the potential to spread resources and employment more widely.
- Biomimicry refers to a revision of the current "take–make–waste" scenario of resource use to "borrow–use–return," echoing natural processes and reducing toxicity.
- Transition to economic activity that enhances quality of human well-being versus a current prioritization of quantity can be achieved by increasing the delivery of services instead of products.
- Jobs can be created, human well-being advanced, resources replenished, and the future made more secure by investing in sustaining and restoring natural capital.

Innovations in technology may alleviate some current constraints, yet many resources and ecosystem services are not substitutable through technology, at least, not yet. Stable weather patterns, fresh water supplies, and arable soils that produce foodstuffs or forage for livestock are the bare minimum for survival, and studies indicate they are all in jeopardy.

Living systems cycle air, water, and nutrients. The health of these systems is declining. Biodiversity, which provides ecosystem resilience, has evolved over hundreds of thousands, and millions of years. Such complex systems, which are not yet fully understood, are not replaceable within a few decades. The risk in upsetting this delicately balanced set of arrangements is immense, and its ramifications will likely affect everyone, everywhere.

A frequently used business aphorism originally uttered by John F. Kennedy, and echoed by Nixon, Gore, and others over the years, speaks especially to the current wind of change toward sustainability: "When written in Chinese the word crisis is composed of two characters. One represents danger, and the other represents opportunity."[8] While this interpretation of the Chinese symbol has been challenged, the concept of equating risk with opportunity has established roots in commerce. Social and environmental change will have an impact on your company, if it has not already. What strategic objective could be more compelling and urgent than addressing the risks and opportunities created by these circumstances?

The Myth of Externalization

The biosphere, then, might be thought of as the envelope within which all social and economic activity occurs. That doesn't mean the environment is prioritized above them, only that social and economic processes and policies must be aligned with the viability of global, regional, and local ecosystems rather than disregard them. The aim of the economic system and its subsystem of industry cannot work at cross-purposes to the larger biospheric system on which it depends. Similarly, economic and environmental processes and policies will not work in the long run if they operate at odds with social health.

The term that is used to identify the disconnect between economic prioritization and social and ecological viability is "negative externalization." The term refers to the costs associated with economic activity that are externalized, or not accounted for by the firm in their operating costs or price structure, but instead imposed, knowingly or unknowingly on parties who have not agreed to these costs. These are the costs mentioned earlier, such as public safety hazards, pollution, and resource depletion.

A number of these costs have started coming back to roost:

- In the late 1980s, Exxon made headlines with the Exxon Valdez oil spill, costing the firm over $3 billion in cleanup costs, fines, and compensation.

- The costs to BP for the 2010 Deepwater Horizon spill are likely to dwarf that.
- Nike experienced a scandal over child labor and sweatshop suppliers in the 1990s, damaging its global reputation, motivating it to improve its practices, and attend to these and other labor issues in their supply chain. They are still working through these problems more than 20 years later.
- In 2012, Apple had its "Nike Moment" incurring a prolonged spate of negative publicity over the labor practices of one of their suppliers, Foxconn, one of the world's largest electronics parts suppliers. Forced overtime, underpayment of wages, manufacturing health hazards, and safety issues were among the problems cited.

Methods to work GHG emissions into commerce, such as a cap and trade scheme or a carbon tax, have not yet been successful but many experts believe it is only a matter of time before the capture of these costs into business operations is mandated.

Some of these errors in assumptions, discussed above, may already be obvious. That still doesn't make the adoption of sustainability practices less daunting, especially with a number of myths circulating about CSR and sustainability business efforts. This next section debunks a number of those falsehoods.

Top CSR/Sustainability Myths

"It's Expensive"

Actually, the opposite is most often true. Much of the low-hanging fruit of sustainability initiatives ties directly to expense reduction. Reducing energy costs through improving efficiencies in heating, air conditioning, lighting, transportation fuel costs, and manufacturing processes adds to the bottom line. Retooling production to minimize inputs and maximize resource use lowers the cost of raw materials. Recycling, reuse, and other waste diversion techniques lessen disposal costs. Engaging employees in sustainability efforts increases their engagement in the firm, and enhances attraction and retention, improving productivity and reducing turnover costs.

"It's Too Intangible"

There are a host of quantifiable metrics, as mentioned in the paragraph above. Even so, there are those intangibles, but just about every firm values competitive advantage, brand image, and reputation, and these are intangibles that must be managed. One way to manage them is to find relevant ways to measure, baseline, and monitor them. Practices toward sustainability can be, and often are, designed to do exactly that. Customer and employee surveys and market research designed to capture such information can identify trends, as well as the some of the specifics responsible for the change.

"We Have Other Strategic Objectives to Attend to"

It's likely that sustainability initiatives will help you achieve your objectives and may directly tie in to key performance indicators. In essence, incorporating sustainability practices into your business operations and, further, into your business model, is about seizing opportunities and lowering risks. As will be discussed in the next section, these practices aren't about add-ons but about doing things differently, and better, with quantifiable results.

Building Sustainability into Business Processes

As many forces are driving the internalization of costs that were previously externalized, firms are finding ways to integrate them, while lowering costs, reducing risks, or both, through shifts in business processes. In May 2012, Microsoft announced a commitment to achieving carbon neutrality in their energy use; this means a net zero contribution of GHGs for the energy use of their operations in over 100 countries.[9] To achieve this goal, they are implementing an incentive, a carbon fee chargeback that will be assessed to each business unit in addition to their energy costs. Improving efficiency, incorporating renewable energy, and retrofitting buildings are among the strategies that will be used.

Around that same time, Whole Foods became the first major North American retail food chain to sell only seafood harvested in a sustainable manner. The fraudulent mislabeling of seafood by some distributors was reported earlier in the year, documenting through DNA analysis that the majority of types of fish purchased from restaurants, sushi bars, and grocery stores for the study were not accurately labeled.[10] Whole Foods, however, recognized their risks and worked with the Marine Stewardship Council to participate in its Certified Sustainable Seafood program.[11] The food retailer sources its seafood directly from fishing companies and processes it themselves, thereby making it easier to track and verify the source and species of the seafood it offers. This strategy results in no net negative impact, by their own company, on the future supply of their product.

The Peabody Hotel chain has recently added post-consumer food waste to their organic waste composting program, which was already removing tens of tons of pre-consumer food waste from their waste stream.[12] This relatively small-scale program is an example of investment in restoring natural resources, in this case, soil.

Building incentive into pay, from the front-line employee to the CEO, is one of the process strategies that Intel uses to motivate toward their sustainability initiatives through a bonus plan that links those bonuses to operational goals. Such a strategy serves to link economic benefit for the company to corporate action to reduce the firm's environmental and indirect public safety impact while also tying in to the economic health and engagement level of all employees.

Building Sustainability into Business Models

There is no best way to incorporate practices promoting sustainability into a firm. Every business can increase efficiencies, reduce waste and expense, and innovate its operations. Where the sustainability muscle really meets the bone, though, is through re-examining a company's business model so as to integrate sustainability into all processes, rather than as an add-on afterthought.

Most companies can push sustainability further down into their operations so that it is evident in their business model. Interface, a carpet manufacturer, is an often cited exemplar. The firm was started in 1973 with the idea to offer an improvement on broadloom carpeting: free laying replaceable modular carpet tile.

Interface's founder, Ray Anderson, "got green" in 1994 after reading Paul Hawken's *The Ecology of Commerce*, and made sustainability a strategic priority.[13] In addition to reworking, over time, all of their processes so as to achieve a net zero negative environmental impact by 2020, they also segued from a product business model to a service business model. Instead of selling carpet outright, their commercial division leases it to their customers. When its useful life is over, the carpet is returned to Interface for recycling, removing it from the waste stream.

Cat Reman, the parts division of Caterpillar, charges a deposit as an incentive for customers to return equipment for remanufacture. A number of utilities are changing their business model from selling energy units to providing energy services. Zipcar, a car-sharing model, reduces the need for car ownership or rentals, also eliminating the resources used in the making of what might otherwise be privately owned vehicles.

Business model change might be in the area of value proposition, such as the way revenue is generated, how segments are targeted, or the type of products and services that are offered. Changed models might, instead, or in addition, be based on operating models: retooling the cost and asset model, reconfiguring the value chain, or deploying organizational change strategies to drive competitive advantage.[14]

There are other means beyond a reworking of existing policies, processes, structures, and operations. Some companies have gone further in integrating sustainability into commerce. Social enterprise organizations are organized around providing commercial answers to social issues.[15] Benefit corporations, known as B Corporations, solve environmental social problems by committing to higher legal accountability standards, meeting comprehensive and transparent environmental, governance, and social standards, and support public policies by building a supportive business constituency.[16] See Chapter 5 for more on these forms of enterprise.

Patagonia, the outdoor clothing company and a certified B Corporation, is frequently mentioned as a leader in sustainability practices. Among the first to offer organic cotton in their clothing, even going so far to work with their suppliers in this effort, they have integrated many other practices into their supply, manufacturing, distribution, and administration. They have designed policies to reduce raw material use, toxic waste and energy use, cooperate with regional conservation efforts, and even educate consumers. They have even initiated a "buy less" promotional campaign that features a partner agreement with eBay that provides an internet portal so that customers may sell their used clothing. As radical as this appears, they still expect to maintain financial health by raising prices, selling products to a wider sustainability-oriented consumer base, and expanding into new categories.

Environmental Qualities of Sustainability

There is progress at building momentum around sustainability on the micro level. More companies are issuing sustainability reports, consumers are becoming more aware and educated about sustainable consumption strategies, and governments are starting to develop sustainability departments. Yet, on the macro level, just about all the indicators continue to go the wrong way: GHG emissions, deforestation, soil loss, species extinctions, and energy and water use continue to climb. As serious as these problems are, and they are the most urgent and important issues of our time, all of them harbor a host of possible profitable ventures with massive payoffs.

Air

Of course, air itself is abundant, but clean air is another matter. The ramification of particulates in the air from sources of pollution includes health impacts, alteration of rainfall patterns, reduced visibility, and reduction of sunlight. The WHO reported in 2011 that the air quality of a great majority of urban areas around the globe exceeds their recommended guidelines. The WHO's research estimates that deaths worldwide due to breathing particulate matter are over 2 million each year.

Coal plants are certainly not the only culprits: industrial processes and transport, as well as personal transport and biomass burning for cooking and heating use, along with biomass burning for agricultural purposes contribute to the decline in air quality. The loss of human life is compounded by the economic cost of related health expenditures (estimated to be in billions of dollars in Europe alone).[17]

Ensuring clean air for use in business operations, particularly in and near urban areas, is getting harder to accomplish. Even if vigorous steps are taken, firms are at risk of increased expenses, and not only for facilities management to filter air or fleet management costs to reduce exhaust emissions. Health and insurance costs are likely to increase, while at the same time employee productivity may decline from respiratory ailments. Insurers are likely to begin requiring their policyholders to address these issues or face steeply rising premiums or policy cancellations.

Technologies and processes to reduce particulate matter during combustion, radically increase efficiency of fuels, and to scrub emissions are among the opportunities in this sphere.

Water

With over 70% of the earth's surface covered by water, you would reasonably think that there is more than enough to go around, but only 2.5% is fresh water. Of all water resources, 97.5% is contained in the oceans. It would be great if climate-change-related sea level rise could be resolved by using more ocean water, but the amount of energy and the expense in desalinization has historically prohibited its widespread use. The search for more efficient and affordable desalinization processes is growing and represents a huge business opportunity, especially as meteorological forecasts predict more frequent and widespread drought, notably in areas that are already arid.[18]

How critical are fresh water supplies to your firm? Consider the risks it would face through an extended drought-induced water shortage, such as occurred in China, East Africa, and areas of the United States in 2011 and throughout the U.S. Midwest, Russia, and the Ukraine in 2012. While agricultural firms and utilities are among those most obviously at risk, manufacturing and operational processes are likely to face higher costs and constraints on water use.

Wastewater represents both a management problem and a solution to fresh water resource constraints. The treatment of wastewater for reclaimed use is becoming more widespread, particularly as drier conditions prompt water utilities, cities, and municipalities to take action to reduce risks to the public water supply. Businesses that provide products and services related to water conservation and reclamation practices and technologies that hold promise to increase efficiencies have a variety of markets they might operate in, from the utility and local government level, to business clients, and to consumers.

Soil

The thin crust of soil that sits atop the surface of landmasses may seem like just dirt, but it provides the nutrition, anchoring, and moisture crucial for plants, upon which all life, including human, depends. Like a skin covering the earth, it creates a boundary that also absorbs and breaks down compounds in air, water, plant material, and minerals. Aside from its service as the source from which all nutrition is derived, soil performs another function: it sequesters carbon from the atmosphere, which is then taken up by plants. The skin peeled off one quarter of a quarter (1/16) of an apple would approximate the comparative percentage of earth that comprises soil.

Soil erosion is a natural process caused by wind and water, but clear-cutting, mountaintop removal, modern agricultural tillage practices, land development, and other human influences have accelerated soil loss. The term for the result of soil loss, rendering land unable to hold moisture, provide nutrients, and create a stable base for plant life, is "desertification." Close to three quarters of the world's rangelands have been degraded by desertification, and over the last 20 years the amount of soil that eroded amounted to enough to cover the entire cropland of the United States.[19] It takes 1000 years for one inch of soil to form, but only seconds for it to be eroded.

Fortunately, there are many techniques for conserving soil: tilling, cover-cropping, and land contouring practices, selective cutting, amendment of soil with organic matter, permacultural practices, and smarter land development, are among them. They are not yet widely practiced, however, so there is room for substantial growth possibilities for providers of services and products that conserve or prevent the erosion of soil.

Oceans

Fisheries provide about 16% of the world's animal protein, with higher percentages in industrialized nations.[20] As climate change is predicted to impact agriculture negatively in many locations, dependence on fisheries may grow in importance. Seafood is currently harvested at a rate that has resulted in the decline of 13 of the 15 major global fisheries, with a steadily declining catch, in spite of one-third of fish stocks being fully or overexploited. In the process of obtaining seafood, over one-third of the total amount caught is bycatch, or sea life that is thrown back, increasing the mortality of other ocean creatures that are potential future food stocks, and reducing ocean diversity and resilience.

Pollutants and waste from industry and human consumption, and rising acidity levels due to dissolved carbon dioxide and declining oxygen levels due to increased temperatures have been documented as profound threats to ocean life.[21] Phytoplankton in the oceans generate fully half of all oxygen in the atmosphere, and are the base of the food chain for most of the fish consumed, as well as much other marine life. But the population of phytoplankton has dropped by 40% in the last 60 years.[22]

The fishing, seafood, and tourism industries have already felt sharp impacts from these developments, but all businesses will have to contend with the effects of lower oxygen levels and higher carbon dioxide levels in the atmosphere as well as the oceans. Already three U.S. states, California, New York, and Washington State, had begun to mandate that insurers disclose climate-change response risks and plans. Insurance companies not yet requiring their policy holders to address climate-change risk and action plans are likely to soon begin doing so.

As with all problems, opportunities abound. Services and products that aim to increase the health and future output of fisheries, that reduce bycatch, that find and market other types of marine products, and that monitor and measure factors affecting ocean health are business gaps in the current future of this resource.

Another influence upon the oceans that impacts human economic (and all other) activity is its contribution to weather. Research shows that changes in temperature influence surface and deep ocean currents, which then influence atmospheric air and precipitation patterns, leading to short- and long-term weather changes. See *Climate*, below, for more on this connection.

Climate

One or two hot spells do not equal climate change. Weather extremes can and do happen from year to year across the globe. Climate change is judged on more rigorous and extended observations. Meteorologists have, unfortunately, come to that conclusion, based on the increase in frequency and severity of extreme weather events. Tornadic activity, droughts, floods, and heat waves, as well as the incidence of disease among human and livestock populations, and insect infestation impacting agricultural output are causing steep increases in economic losses, not to mention human suffering.[23]

Added to the risks mentioned in the discussions of other resources above, the effects of climate change hold the greatest potential for infrastructure breakdown, regional and national economic blows, and social upheaval. There is, arguably, the largest window of opportunity for businesses that provide services and products for governments, businesses, and individuals to mitigate and adapt to climate change.

Biodiversity

The number and range of life-forms, including animals, plants, and microorganisms on earth is estimated to be over 100 million, yet 91% of marine species and 86% of land species have yet to be discovered and identified, according to the journal *Nature*.[24] Biodiversity is important for many reasons, the broadest being the ability to rebound after damaging impacts. A diverse biological environment provides ecosystem, biological, and social services. The smaller species aid in soil formation, nutrient recycling and storage, breakdown and absorption of pollutants, and water cleansing, among other services.

The fishing industry offers an extensive range of materials and food stocks, each of which needs a wide enough genetic diversity to keep its populations viable. As some of these resources are being depleted, their gene pools are shrinking, leaving stocks that are more vulnerable to diseases and other environmental pressures that could, and apparently are, causing their numbers to crash. A research study in 2011 that concluded that species extinction rates are overestimated still cautioned that species extinction was a very real and present threat that required attention, and that was likely to accelerate quickly.[25]

Extinctions represent not only the loss of a life-form that has been millions of years in the making, but fit into an environmental niche that is largely not yet understood. Without that understanding it is impossible to know how significant those losses are, nor how severely the losses will affect the vigor of the systems within which they are interdependent. Due to the complexity of biospheric systems, science is as yet, and perhaps will never be, able to understand no less replicate or replace these losses. Even if the costs could be calculated, they are likely to be incredibly expensive due to the interdependent nature of ecosystems and their subsystems.

Here, too, there are many places for commerce to intervene. Ventures such as ecotourism, organic agriculture, and sustainably harvested timber are already experiencing growth despite a shrinkage in overall global economic vitality. Mitigation and monitoring services, technical services, innovative financing, and bioprospecting are other avenues for growth.

While the threats to environmental health are serious and numerous, businesses are in the best position to address them. Firms that ignore these perils set themselves at a competitive disadvantage both in the short term and in regard to their longevity.

Economy and Social Sustainability: Conceptions of Wealth

Recalling a point made earlier, when employing a systems perspective, it is crucial that any subsystem operates within its enveloping system's purposes and constraints. Stepping back from commerce to take a broader view, this section examines the economic system within which it functions. Noted American economist and Columbia University professor Joseph Stiglitz said about the purpose of the economy: "It's not to produce GDP. Let me make that clear. The purpose of an economy is not producing GDP. It's increasing the welfare of citizens, and it's increasing the welfare of most citizens."[26]

The global economic system is currently premised on material wealth. The concept of wealth is typically considered to be a state of economic abundance. But there are other aspects to wealth, aspects that actually describe why people seek to accumulate it in financial form. Aside from providing for basic needs, financial wealth may—but not always does—afford an acceptable state of health, an adequate sense of security, access to education, a satisfying family and social life, involvement and voice in one's community, and sense of purpose. In a broader sense, financial wealth is pursued because it is believed to increase levels of freedom and happiness.

For those in the developed world, beyond a moderate level of wealth, additional income and added material comfort do not necessarily equate with positive feelings. A large multinational study, surveying over 130,000 individuals in 132 countries, evaluated income against a set of measures of well-being. Although income was related to life evaluation, it was not related to positive and negative feelings of social psychological prosperity.[27]

Other studies have shown that even with rising incomes, measured by real gross domestic product (GDP), happiness levels within most countries, including the United States, have remained flat over a period of decades. Well-being increases significantly at lower income levels, but greater and greater levels of income do little to improve well-being.[28]

Research has also been conducted to evaluate measures of well-being with material flows. One such study measured human well-being as life expectancy, flows of physical capital as GDP per capita, flows of natural capital as the ecological footprint, and human capital as education. This study assessed data from 135 countries and found that human well-being was not increased through the exploitation of environmental capital (consumption).[29]

Consumer spending is the largest component of GDP, although it also measures net exports, government spending, and investments. As a metric for gauging national economic health, it reduces monetary activity to a simple quantitative answer, separating it from, and potentially leading to conflict with, the larger system within which it operates.

Enterprise at the Base of the Pyramid

As confusing as the statistics are, while the global middle class is mushrooming it is also estimated that half of the world's population lives on less than $2.50 per day and over 80% live on less than $10 per day. The number of children worldwide living in poverty represents one of every two children on earth. The ratio of those in wealth to those in poverty has steadily increased from 1 in 11 in 1913 to 1 in 35 in 1950 and to 1 in 72 by 1992.[30] As sobering as these facts are, they point to some of the enterprise activities being recognized and mined by savvy business organizations.

The bottom of the market is known as base-of-the-pyramid (BOP), meaning that it occupies the lower end of a figurative socioeconomic pyramid. Businesses designed to reach acceptable commercial rates of return and scalability are providing basic services such as access to water, education, and healthcare, sometimes partnering with government agencies or NGOs. Organizations such as Grameen Bank, a for-profit antipoverty bank, and Sanasa Development Bank are among a growing slate of microfinance institutions serving this market.[31] Those in the BOP also represent a pool of largely untapped entrepreneurial talent, labor, productivity, and innovative distribution systems.

The private sector has barely explored the BOP to date, apparently having decided that this market is not lucrative enough, or perhaps too risky, and leaving it to development organizations to address. Some needs being filled at the BOP by enterprises are rural electrification, health and medical products and services, water extraction and purification, food production, communications, transportation, and housing.

Chapter Summary: Key Takeaways

This chapter delved deeper into the TBL aspects of sustainability, employing a systemic perspective to explore existing economic, environmental, and social circumstances. It began by reviewing evidence, or the lack of evidence, supporting a number of economic assumptions, which are based on a worldview that is less sufficient in explaining facts, and for which a systemic explanation and the long view of sustainability make a lot more sense.

These flawed assumptions include that growth benefits everyone, that people make decisions rationally and only for their own benefit, that unceasing continual growth is possible, and that costs for business impacts not directly incurred by a firm are outside their responsibility.

Instead, research shows that growth benefits few, and that it fuels growing income inequity, that emotion drives decision making as much or more than logic and cooperation is as potent a force as competition, that constrained resources are forcing disruptive innovation in business, and that commercial interests can no longer externalize their impacts on society and the natural environment.

Similarly debunked were commonly held assumptions about CSR and sustainability in the business context; that it's too expensive, too intangible, and not important enough to pursue. Sustainability efforts can and often do lower expenses, are tied to solid metrics, and will boost any short- and long-term strategic plans.

In the global context, sustainability must address well-researched and documented crises within the environment and society. Industry is among the most significant drivers of the degradation of air, water, soil, oceans, climate, and biodiversity, all vital natural capital at considerable risk of becoming ever more scarce, expensive, and insecure.

At the same time, these problems offer solutions for enterprises to address a subject covered further in the next chapter. A systemic examination of the social side of sustainability reveals that economic and environmental issues are related to issues of poverty: access to clean water, food, energy, education, healthcare, and livelihoods are all areas around which enterprises can build new businesses and models.

CHAPTER 4

Commerce as a System

The previous chapter examined sustainability as a system in terms of social, environmental, and economic impacts and described how systems thinking could reveal to business leaders opportunities to more efficiently use both human and natural capital. Risks and rewards inherent in issues such as environmental pollution, water use, and unrestrained growth were analyzed (the next few chapters expand further on these rewards and risks). When looking at these issues through a systems lens, managers can design innovative strategies to capitalize on openings for generative, restorative, or risk mitigation services and products.

This chapter will look at commerce as a system, beginning with the extraction/harvesting phase, going through production, distribution, consumption, and disposal. Examples of real-world companies illustrate the points made in each section. While the focus may appear to be on manufacturing, service businesses also significantly impact sustainability issues. Any facilities and operations, including banking and financial services, retail, and other service industries, even if they appear to be mainly administrative in nature, have negative footprints.

The Interdependent Components of Commerce

As with most terms, commerce has a variety of definitions. From the individual to the multinational, and from the sharing of ideas and opinions and, of course, goods and services, commerce denotes an exchange of something between parties. Commerce is usually thought of as a process in which a means is utilized to achieve an end such as enriching oneself, one's company or state through trade. Examining commercial enterprise more closely, though, much of its benefit occurs during the process rather than as a result of it. There is often engaging and broadening social and cultural interaction between individuals and groups during an exchange.

The buyer gets satisfaction from getting a need filled and the seller experiences the same in filling that need through recognition of the value of what is sold or exchanged. Relationships are forged and strengthened.

Beyond the human component, though, commerce can be viewed through a wider systemic lens. Although commerce may be thought of as a unique human activity, exchange is the most fundamental of activities within any living system. Even single-celled organisms exchange gases for the purposes of respiration, trading the carbon dioxide their metabolism produces for oxygen in the air that passes through their membranes; their sub-cellular structures break down carbohydrates and fats for the needed energy.

Economic commerce, echoing this most basic of transactions, is the reflection of a much larger system within which it operates. Materials are extracted, processed, converted, and exchanged. In the larger system, one entity's waste is another's resource. Existence is ensured only if certain conditions are met. The materials to be extracted must continue to be available, and without more labor (read expense) than the entity can afford to exert. The processing function must operate adequately and in concert with the entity's purpose. There must be entities with which or whom to exchange all byproducts, and these entities must be viable enough to accept and afford the exchange. The ecosystem of the entity must be stable enough to support its function, and the functions of its exchanging entities over time.

From an environmental perspective, these ecosystems have developed over hundreds of thousands—in many cases, millions—of years, producing a wide variety of entities that allows the ecosystem to attain a dynamic equilibrium. Gradual change, over thousands or hundreds of thousands of years, allows individual species time to adjust, if conditions support enough of them to maintain a viable population. Sudden change puts intense pressure on ecosystems to adapt or die. Such rapid change may result in conditions that allow some entities to prosper while others wither and disappear if they cannot adapt, but there is always an interdependence of entities upon one another. Just as an ecosystem cannot survive if there are too few adequately functioning entities to allow it to survive, a business must have

- adequate, appropriate, and affordable materials;
- willing and able suppliers;

- effective processes;
- a productive and healthy labor force;
- customers who want and can afford their products;
- viable channels through which to reach their customers for communications and distribution;
- vendors or processes to accept or convert waste and byproducts.

For Service Industries

The phases of commerce as discussed later in this chapter are more applicable to product-oriented firms than to service firms, but sustainability is also needed in the latter. In these industries, the effort is less about environmental impact reduction than attending to their social and governance issues.

A focus on energy, water, and waste is somewhat less material to the operation of service firms than it is to manufacturing firms, although real estate assets and facilities do have significant environmental and social footprints. That's not to imply that service firms should ignore the environmental leg of the sustainability platform, but that green claims, especially at the expense of social and governance issues, may be perceived by onlookers as naïve and shallow or, worse, manipulative and deceptive. Instead, areas that are more relevant for service firms include

- customer security and privacy;
- customer transparency and fair dealing;
- responsible outsourcing;
- talent recruitment and retention;
- employee compensation and working conditions;
- workforce diversity;
- financial inclusion;
- regulatory compliance;
- anti-corruption efforts;
- fair market competition; and
- justification of government subsidies or bailouts.

There are opportunities for service firms to integrate environmental aspects deeper into their business model than simply attending to their buildings and administration. For example, financial service firms might offer green securities, such as climate bonds or alternative energy investment options, loans that incentivize energy efficiency in residential home and business facilities, and credit services that encourage the purchase of options which are more environmentally sensible. Underwriting standards can be improved to require that businesses, real estate, and markets attend to their environmental (as well as social and governance) practices.

Just as product firms do, service firms can also influence their suppliers to comply with codes of conduct that outline governance, social, and environmental standards.

A Systemic Production Paradigm

Beginning with his book, *The Ecology of Commerce*, Paul Hawken explored the existing model of capitalism, further elaborated in his work with Amory and Hunter Lovins in *Natural Capitalism*. They describe the linear approach that industry has used for the last several centuries as "take—make—waste."[1] The authors observe how exponential global population growth, made possible largely due to industrialization's tremendous and rapid use of natural resources, has led to a widespread decline in living systems.

One primary assumption that underlies this traditional approach to commerce is that the accumulation of financial capital is the solitary measure of success, thereby excluding other measures and factors from consideration. A second assumption is the continued availability of raw natural capital as the main asset base for the economic engine, which the authors compare to living off the "principal" of natural assets rather than on their "interest."[2]

They suggest instead a borrow—use—return approach, which acknowledges the reality of the cyclical nature of natural resource production and replenishment. Rather than reducing waste, this approach seeks to entirely eliminate it, emphasizing a closed-loop production model. Fossil-based energy sources, which are a reservoir of solar energy stored over millions of years of geological processes, are replaced by current solar energy sources such as photovoltaic, wind, wave, and solar thermal sources.

The rates of use of renewable and non-renewable resources do not exceed rates of regeneration or development of these resources, and rates of emissions are held below the rate at which the natural environment can absorb the emissions without a decline in viability. Priority is given to investment in the restoration of natural capital rather than only its continued depletion. The economy shifts, over time, to an emphasis on the provision of services rather than of a greater abundance of cheap products.

The following sections of this chapter offer a glimpse into some of the recent and current disconnects, and reconnects, between commerce and sustainability. These phases represent systemic intervention points for capitalizing on solutions and innovative ways of reworking both the means of commerce and the ends. Figure 4.1 represents the phases of a product as it moves from extraction and production through use and disposal.

In the world of commerce, supplies can come from around the entire globe, outsourcing is common, and short product lifecycles can make easy evaluation of supply chains challenging. An eye-opening example of the complexity of supply chains is offered in Dr. Jane Macfarlane's detailed operations illustrations for the production of men's cotton slacks, polyester/cotton bedsheets, and Nylon Supplex parkas in which there are well over 40 process steps.[3]

The slacks illustration process map begins with the field where the cotton is harvested and travels along a road on which each step of the process is documented, from harvesting through production and ending with display at the store where the consumer can purchase the product. These single-page illustrations, as comprehensive as they are, comprise the process steps that make up only the production phase of the commercial cycle.

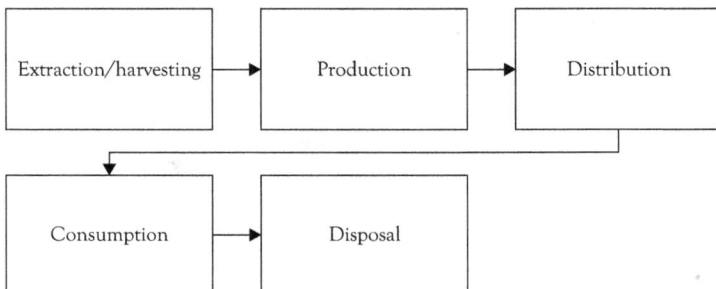

Figure 4.1. Basic production diagram.

Note that while the necklace example examined below utilizes a product in the jewelry industry value chain, service businesses produce collateral materials and other products that follow the basic production path shown in the diagram, for which there are opportunities to improve practices. For the purpose of illustration, this chapter will follow the phases of commerce involved in a necklace of gold chain and turquoise stones, from the extraction of the gold and turquoise to their refinement into chain, clasp, stone shape and condition, to its packaging, distribution, purchase, and disposal. Such a product has an impact on not only the buyers and sellers along the chain, but the environment and the communities in which the raw materials were extracted, where they are processed

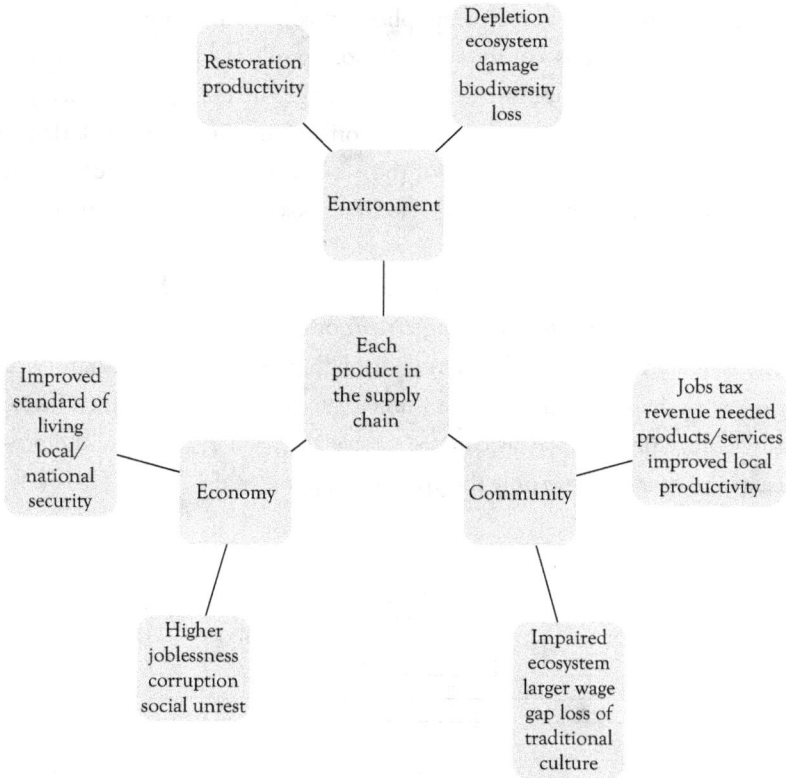

Figure 4.2. Interactions of each product in the supply chain with the environment, community, and economy with examples of positive and negative impacts.

Extraction	Production	Distribution	Consumption	Disposal
Gold	Refinement of raw products	To retail shows	Purchase	Reuse
	Assembly of necklace	To reps		Recycle
Turquoise	Packaging	To markets	Use	Landfill

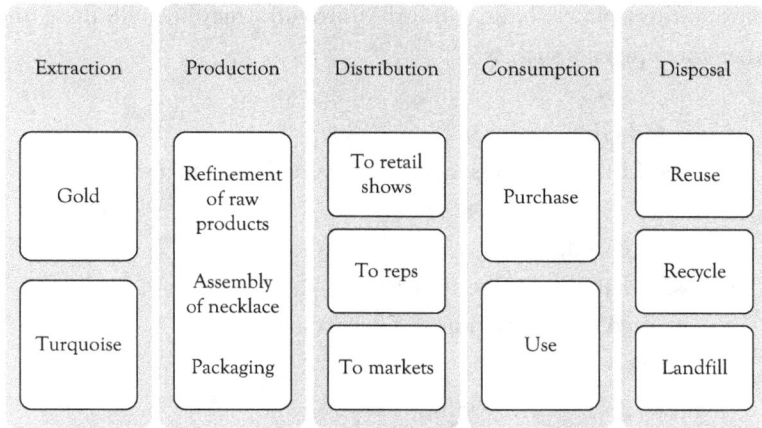

Figure 4.3. Elements included in the production of turquoise necklace.

into usable parts, where the parts are assembled and then distributed, and where they are purchased and ultimately disposed of. Included in this journey is the transportation to get the raw materials and other resources required to make the pieces of the necklace, and the packaging and transportation to get it to the market, all of the labor used in the production, the ultimate purchaser/consumer, and whether it is reused, recycled, or landfilled. Figure 4.2 represents some of the positive and negative aspects of a commercial supply chain, and Figure 4.3 depicts the cycle of commercial phases in more detail.

Extraction and Harvest

Extraction and harvest are the collection and aggregation of raw materials, whether through mining for precious minerals, drilling, and fracking for oil, gas, or water, or harvesting of timber products, plants, and animals for food and other purposes. In the case of the gold and turquoise necklace, some of the environmental impacts include

- land disturbance;
- use of toxic chemicals to extract the stones and gold from the rock;
- air and water impacts from mining operations.

Impacts from materials used in other products may include these and other consequences such as

- air and GHG emissions;
- runoff from fertilizers, herbicides, pesticides, and animal waste;
- toxins in soil;
- loss of habitat and biodiversity;
- impaired water filtration and flow systems;
- and more.

Some of these are regulated to a greater or lesser extent although enforcement is not stringent enough to restrict many incidents of non-compliance. In these instances, policy interventions could affect businesses, such as tighter government enforcement or incentivizing compliance to make it the default choice. In countries where regulations do not provide protection, some forward-thinking firms, especially those from developed countries, have taken matters into their own hands by following consistent protocols that protect the land, water, air, and viability of communities who were living in the extraction area prior to that activity.

Extractive industries have sustainable development potential. They may contribute to employment in remote and depressed areas while also stimulating a country's economic growth, often as first opportunities for private sector development and foreign investment. Yet, they have also caused or aggravated social inequality, ecological damage, health issues, and political corruption.

The extraction of great resource wealth has bred conflict, which has at times escalated into war. The World Bank has reported that "[m]any resource-rich countries perform worse than resource-poor countries in key aspects of development, including economic, social, and governance."[4]

Social impacts on communities may appear to be difficult to assess, but examining quality of life before and after extraction begins is one place to start. Access to basic services, education, medical attention, clean air and water, infrastructure, security, and other indicators can be measured. Doing so permits the extraction company to see areas that can be mitigated.

One area that is not well regulated is animal production for food (confined feed operations, slaughterhouses, and processing plants), not to mention treatment of the animals, public awareness of which has risen recently through the efforts of undercover reporting. Some states in the United States are looking to ban efforts to reveal farm animal abuse through legislation.[5] The working conditions, health issues, and wage levels of workers in these facilities are another consideration.

The International Metals and Mining Council states that in order for mining to be conducted at the highest standard possible, "a deep understanding of mining's contribution to sustainable development is essential...and requires consideration of both the opportunities and benefits achieved through the extraction, processing, and use of minerals and metals, as well as an assessment of the economic, social and environmental costs and risks of doing so. Importantly, the responsibilities of all stakeholders—government, civil society, communities and companies—must be considered."[6] The industry realizes the importance of considering numerous aspects of impact.

Newmont Mining, one of the largest gold and copper mining companies in the world, includes its sustainability commitment in its mission.[7] Its sustainable objectives include building community relationships, ensuring preservation of non-mineral resources, safety, and promoting sustainable economic development.[8] This is a public company in which management has realized the value of considering not only profit but also worker safety, stakeholder impacts, and environmental stewardship.

Newmont Mining company has systems in place to ensure its adherence to these principles, including using local suppliers where possible, community training or workers, and waste management. By looking at all of the relationships involved in its mining operations, Newmont has created a system that examines and attempts to mitigate all adverse effects to the environment, the community, and its workforce.

Loss of traditional subsistence living due to mining pollution creates conflict and forces indigenous people into deeper poverty. In the Niger Delta, oil has been extracted since 1958. Fish ponds have been poisoned due to the mining activity causing the fishermen to lose an important food and income source.[9] These negative externalities are not considered a cost to the oil company, Shell. The stakeholders in this region are adversely affected with little recourse. Although Shell was forced to pay a $15 million dollar fine, the subsistence living of the people in the area will not be replaced. As recently as 2011, a report issued by a division of the United Nations stated that, "natural resource extraction and other major development projects in or near indigenous territories are one of the most significant sources of abuse of the rights of indigenous peoples worldwide."[10]

Some believe that in the building of economies in developing countries allowances should be made, the sacrifice of a few current citizens to benefit the well-being of a greater number of future citizens. They suggest that safety protocols, restorative compensation, and the like are too time consuming and expensive in the race to be better. Whether those whose lives, livelihoods, or well-being are sacrificed would agree, or whether consumers in developed countries reaping the fruits of those sacrifices would still purchase those products knowingly are questions firms might want to ask.

"Conflict minerals" is the term applied to the raw materials that are extracted and sold to finance militias and gangs, armed groups that trade illicitly through smuggling schemes. Diamonds extracted in war zones in trade for arms or to fund insurgencies or invading armies in Angola, Liberia, Sierra Leone, and other African countries have led to international agreements to halt such trade. Gold, tin, tungsten, and other minerals have been sourced in this manner from Congo, acting as the base of the supply chain for large well-known electronics manufacturers. Media attention has brought these issues to light, and they are now being resolved through legislation and supply chain certification programs.[11]

Long the primary global supplier of rare earth minerals, a collection of elements vital to the production of modern electronics, China holds almost a quarter of the global available quantity. A white paper recently released by the Chinese government reported that two-thirds of the country's supply of these resources has already been mined and that the balance consists of much poorer quality seams, which would increase the expense of extraction.[12] Ramped up production in the 1980s and 1990s created an oversupply, which drove down prices, and illegal mining then added to the oversupply problem. These mining practices also had damaging environmental and social impacts, flooding, landslides, and deforestation among them. China's rare earth mining industry is an example of unexamined and unsustainable extraction policies and processes that quickly led to an imploding materials base and the seizing up of this particular engine of economic development.

During extraction and all phases of a product's lifecycle, resources are often wasted. This could be a behavioral issue like wasting energy by leaving trucks idling or leaving computers on overnight. It may also be an infrastructure issue such as having a building that is not insulated well or has a heating/cooling system that is not properly designed. It could also be a quality control issue such as using too many chemicals to extract a mineral resulting in more downstream pollution or improperly installing or monitoring an oil well, such as BP's Deepwater Horizon. The result is wasted resources. Clear-cutting a forest illustrates a lack of a systemic understanding of how runoff will impact the ecosystem and surrounding community, and will result in soil loss, removing vital nutrients for the next generation of forest product. Waste is produced in each phase of commerce, an area ripe for cost savings, revenue production, reputational enhancement, and risk mitigation.

Production

When examining production, a systemic perspective considers all of the relationships and interconnections of the process, its effects on workers, the surrounding community, other stakeholders, and the environment. In a simplified version of a manufacturing facility, the inputs to the factory include trucks delivering supplies and taking away the final product, air and water impacts from the factory itself, and ramifications for the members of the community who live and earn a living in the community. This includes traffic and safety considerations for the employees and the larger community, in addition to air quality, noise levels, and infrastructure impacts. Outputs include not only finished products but wastewater and modified air products.

Gold and turquoise must be refined to the proper size and shape in order to produce the gold necklace. The raw materials must be transported to one or more facilities. Refinery, assembly, and packaging plants process gold ore, cast the refined gold into chain, cut and polish the turquoise, assemble the pieces into the necklace, and package and label it for distribution.

As with the extraction phase, workplace health and safety is a major issue in the production phase, so much so that there are a variety of agencies that address these types of problems stemming from commercial activity. The Occupational Safety and Health Administration (OSHA) is the agency responsible for worker protection, the Mine Safety and Health Administration's mission is to protect mine workers, and the Environmental Protection Agency (EPA) has been instrumental in minimizing exposure of farm workers to pesticides. Yet, there are still widespread problems in ensuring health and safety protection for workers in production.

A few recent examples include the use of some flammable floor finishes and sealers in Massachusetts which resulted in deaths and serious neurological damage. Diacetyl, the butter flavoring used in popcorn, was pinpointed as the cause of disabling lung damage in employees at a food flavoring manufacturing plant in Missouri.[13] Meat and poultry processing workers still suffer from high rates of injury and illness, while plant sanitation and food quality remain problematic.[14] Quite recently, the apparel industry experienced significant negative press after poor regulation and management at production facilities in Bangladesh, the world's second largest apparel exporter, resulted in deaths from fires and a building collapse.[15] These are just a small set of incidents regarding workplace safety issues that exist in the production phase of commerce.

Some industries and jobs are simply more perilous than others. Progress has been made in improving working conditions, usually through standards-based practice supported by regulation and legislation. The most effective change has come from a prevention approach, and from reworking processes so that they are safer to start with. These kinds of redesign efforts represent potential areas for future advancement in many types of businesses.

As companies become more aware of how pollution prevention, resource efficiencies, and tighter operations can be profitable, many are assessing sustainability projects to see where new savings can be found. Studies show profitability when implementing sustainability actions.[16] As a manufacturing company with significant air emissions, 3M instituted their Pollution Prevention Pays Plus program in 1975, well before environmental mandates, to create a company culture that would work to clean up its operations.[17] Currently systems are in place that measure, track, and manage environmental compliance, climate change, energy, air quality, waste, water, biodiversity, product lifecycle, and packaging.[18] Not only have their efforts improved air, water, and waste pollution, the culture of the company allows for innovation at all levels to help prevent pollution and rewards employees for their efforts.

Apple, manufacturer of computers, smart phones, tablets, and electronic accessories, measures its environmental performance by each product since 2009.[19] Measurements include lifecycle analyses, energy efficiency, materials used in production, content of toxic substances, and recycling potential. With respect to recent media reports regarding social impacts through some of its suppliers in southeast Asia in 2012, Apple increased their audits 72% over the prior year.[20]

Sometimes industries receive help from activists. Rainforest Action Network (RAN) has been successful in its work to save the destruction of rainforests for paper and palm oil by leaning on corporations and highlighting travesties occurring in the tropics. In 2011, RAN was successful in partnering with Disney as an ally to stop the production of paper products from indigenous rainforest trees.[21]

The production phase offers a firm a large circle of influence to design and implement change. Small companies may have a limited ability to induce upstream supply chains to alter their processes and practices, yet may still substantially improve their operations. Consumption and disposal phase practices, while historically out of the hands of producers, are changing through design and return/recycle programs. Larger companies have the influence to put pressure on their supply chains to make change. Production has often been the area of focus on the part of media attention and activism in persuading organizations to improve the environmental, social, and economic ramifications of corporate activities.

Distribution

For the purpose of simplification, in this systemic review of commerce the distribution phase has been placed between production and consumption, but in reality, for most finished products, distribution of inputs (and outputs) occurs at many stages through the enterprise cycle. All of these distribution steps incur impacts, and a lifecycle analysis will reveal the significance of those steps. (Lifecycle assessment [LCA] is discussed in more detail in Chapters 6 and 7.)

Distribution requires a significant amount of energy and can in some cases be more energy intensive than producing the product itself. By measuring all of the resources, both human and natural capital, that go into distributing a product companies can determine where most savings can come from, whether it is transport from manufacturing to warehouse, warehousing and storage, packaging, or direct distribution to retailers or consumers.

In the necklace example, the gold necklace is now an assembled product and must be transported to specialty shops that sell jewelry. In the jewelry industry, this process usually involves three steps. These products are usually ordered in small quantities, and the necklaces are delivered to a warehouse where they can be stored and then delivered along with other goods. Sales representatives travel to retailer shows and choose pieces, such as this gold and turquoise necklace, to sell to the stores in their territory. Jewelry ordered then is delivered to specialty retailers or large department stores. The distribution of this product is fairly straightforward. It may need to be repackaged and relabeled but requires no special handling such as temperature control, and it is not perishable, so no express deliveries are necessary. Many products, however, do require special handling.

One small company that has taken sustainability seriously in this respect is New Belgium Brewing, brewer of Fat Tire Ale. The company performed an LCA on a 6-pack of their beer which showed that one-third of the GHG emissions came from the refrigeration of the beer at the retailer—the tail end of the distribution phase. It is harder to require vendors to adopt sustainable practices than suppliers.[22] New Belgium's product must be refrigerated during distribution, as time schedules are important, and therefore requires more forethought in the design of their distribution logistics.

The company recently became 100% employee owned and achieved B corporation certification, allowing the owners flexibility to determine the company's priorities—they are not directed by profits alone (see Chapter 5 for additional information on the B corporation model). Its sustainable practices include using the methane created by their wastewater treatment as energy, diverting almost 95% of their waste, reducing the amount of water needed to make beer from the industry average of 5.0 units of water to 1.0 unit of beer to 3.5:1 by 2015.[23] New Belgium's current distribution originates at the sole manufacturing facility in Colorado, though the beer is distributed nationally.

Another example is Nestlé, a food and beverage producer based in Switzerland, with a much wider distribution net. This company distributes over 125,000 metric tons of food and beverage products worldwide. By looking at many metrics including cost per kilo of product, this company is optimizing distribution networks and route planning, reviewing sea and rail in place of road transport, improving driver training in terms of efficiency and safety, phasing out hydrofluorocarbon, a GHG inventory, and implementing energy saving proposals in the warehouses.[24] Examining and measuring the systems on both large and small scales has allowed Nestlé to reduce its impact on the environment and in the communities where it produces and sells its products.

UPS and Fed Ex are the largest package distribution companies based in the United States. In 2012, UPS received the highest rating by the Carbon Disclosure Project for transparency.[25] Years ago, UPS rerouted its delivery trucks to ensure that drivers only make right turns so as to save time and fuel.[26] This small change not only decreased its fuel consumption, but package delivery became more efficient and thus customers have been better served.

In 2012, FedEx had almost reached its 2020 goal of improving fuel efficiency of its aircraft by 20% (from 2005 levels) and then increased that goal to 30%. Vehicle fuel efficiency has increased by 16.6%. It has added electric vans and even electric tricycles in France to decrease its energy use and GHG emissions. The company also initiated a carbon neutral envelope shipping policy by offsetting emissions. These systematic changes (though many people may not even notice) decreased FedEx's fuel costs and GHG emissions.

Through the encouragement of reusable grocery bags, the Publix supermarket chain now saves over 40 million plastic and paper bags each month and has modified their produce bags to save over 377,000 pounds of plastic annually. The grocer has increased their distribution efficiencies, and decreased their expenses, through adding hybrid and flex-fuel vehicles to their fleet, fitting more into each truck, reducing empty load mileage, and working with truck manufacturers to increase fuel mileage.[27]

These companies use a systematic method of measuring and analyzing their operations to uncover ways to improve resource efficiency and accountability to stakeholders. This is evident across their operations, not only in distribution, but these examples highlight their distribution strategies. Intervention points can be found across the spectrum of a business. Fuel efficiency can be achieved by upgrading vehicles, optimizing delivery and route selection, and training drivers to hyper-mile (use driving techniques designed to improve fuel efficiency). Examining operations within warehouses by, for example, soliciting employees for input, can often save resources. Improving lighting can save electricity and create more comfortable working conditions. Working with employees and communities surrounding facilities and along routes can produce further savings and sometimes good will within the community.

Consumption

Consumers are not well informed as to choices they can make to drive sustainable production and consumption, nor are they incentivized. Supply chains are too complex to understand in the aisle of a supermarket or clothing store. If it costs more, most people are unwilling or unable to pay the premium.[28] A systemic and thorough review of the consumption phase and its connection with the economy, society, and the environment reveals intervention points to decrease negative and increase positive externalities on all three systems.

Currently, the world is accumulating environmental debt, costs associated with restoring ecosystems to their level of functionality before they incurred human-induced environmental impacts. Until restoration occurs, environmental debt rises and is left to future generations, rendering most current consumption levels unsustainable.[29]

Investors have been increasing their sustainability leverage by supporting actions in proxy voting.[30] Although these proposals are not successful in the majority of cases (see Chapter 7 for an example), the trend is clear that investors are paying attention and putting pressure on management to implement social, environmental, and governance practices.

One reason U.S. firms are slow to approach sustainability, unlike many other nations, is that there are currently no federal carbon restrictions. Yet, there is progress: if Walmart is bringing sustainability into its operations, it must be saving money. Its three sustainability goals are "to be powered by 100 percent renewable energy, to be zero waste, and to provide products for its customers that sustain people and the environment."[31]

Other corporations are also on this track because this makes business sense. The sustainability reports of many firms, whether small or large, whether a separate or integrated report, cover resource management and stakeholder assessment, and the savings that come from implementing these processes are often in the millions of dollars. IBM has been cutting its electricity use and expects reductions of 20% from 2008 to 2013.[32]

Every industry is exploring these opportunities. Merck, the pharmaceutical giant, is transparent about its sustainability results.[33] It provides a spreadsheet on its website that shows data from the last 3 years (2009–2011) that indicates energy and water use data as well as environmental health and safety data, including fines. Merck has reduced water use by 10% from 2010 to 2011. Energy use is down by more than 3%. In 2011, Merck was fined 15 times with penalties over $1.7 million; its transparency concerning this issue is commendable as it informs investors, insurers, and consumers of its safety record.

Sustainability is a core value of Yvon Chouinard, CEO of sportswear brand Patagonia. The Footprint Chronicles detail sourcing of all of Patagonia's products with maps of where they are produced, the name of the supplier, and number and gender of workers.[34] Specific suppliers, such as Gortex, offer additional information. Recently Patagonia updated its protocol to include animal sourcing and treatment in products that include materials derived from them. This empowers consumers to make informed choices. (On the distribution side, in 2011, Patagonia changed its port delivery location from Los Angeles to Oakland, closer to its warehouse in Reno, NV, thereby reducing its carbon footprint by 31%, saving $324,000 in transportation costs.[35])

Consumers are in the process of becoming more aware and savvy about the impact of their choices on the marketplace, and how those decisions will drive sustainability. Informed purchasing is already nudging companies to attend to mitigation of their adverse impacts to the environment and stakeholders.

Disposal

At the end of a product's life, it must be disposed of. Once a product is purchased, it will be used by the consumer until it is no longer wanted or needed or has reached the end of its useful life. Options include reuse or repurposing of the product if it is still functional, recycling, or landfilling. In the case of the gold necklace, unless the chain or clasp breaks or the turquoise stone falls out, the necklace can be reused by someone else. If the chain breaks, the necklace can be repaired, or the gold can be sold for recycling and a new chain purchased. This simple example has no significant waste disposal issues.

In the United States, the 2010 per capita waste generation was 4.43 pounds per day with 1.31 pounds of that being recycled or composted.[36] Approximately 54% of our waste stream is discarded, 34% recycled, and 12% burned in waste to energy plants. Metals are recycled 35% of the time, yard waste 58%, newspapers 71%, and glass 33%. A great deal of our waste is discarded in landfills. The decomposition of that waste emits methane, a gas that contributes 20 times more to climate change than carbon dioxide.

As in the earlier phases of commerce, the end of life for a product has its systemic impacts. Economically, waste hauling is an expense. Recycling has the potential to produce a revenue stream because this category of waste can be remanufactured, and most companies could see a cost savings in reuse and recycling over landfilling.

While communities housing landfills usually do receive some financial compensation, impacts include potential safety hazards from large trucks operating in residential or rural neighborhoods, water, air, and soil pollution from landfill contents, and offensive odors from transportation and disposal of the waste. A landfill can be thought of as a system that has inputs (waste) and outputs (methane gas, potential air, water, and soil pollution). Methane gas is a recoverable energy source, a possible revenue producer. Understanding the systemic nature of landfills has resulted in identifying intervention points: prevent leakage from the landfill by lining it and monitoring both downstream water quality and soil quality; implementing waste gas to energy projects to mitigate the methane contamination from the landfill; improving fuel efficiencies of the trucks that are delivering the waste so as to minimize air pollution.

Utilizing ideas from employees for minimizing environmental and social impacts helps improve employee morale (and turnover) and produce sizable gains in efficiencies. Waste Management, employing over 45,000 people, has rebranded itself from a waste disposal company to an environmental solutions company.[37] It manages landfills, converts the methane gas to energy at more than 120 sites and claims to be North America's largest recycler managing more than 8 million tons.

Another intervention point is on the regulatory side. If a carbon tax were levied, even more incentive would be generated as the production of landfill methane gases would be incurred as a cost to the operator, passed along to communities and businesses producing waste. Creation of waste to energy not only mitigates GHG emissions, it would lower cost and increase revenue production.

Some companies like BMW design their products with recycling in mind. Since 1990, BMW has been taking back its old cars for recycling and reprocessing all fluids and solid parts.[38] The design includes fewer materials and recyclable synthetics that allow disassembly and recycling.

Subaru, Toyota, General Motors (GM), and Honda have zero waste manufacturing plants in the United States and Ford has pledged to reduce waste on car manufacturing as well.[39]

By examining all inputs that are used to make a product, a company can find ways to design recyclable or inert components, take back used products, and otherwise mitigate disposal impacts.

Chapter Summary: Key Takeaways

There is significant overlap between extraction, production, distribution, consumption, and disposal. Sustainable products include those that utilize fewer resources both in their manufacture and use, that last longer, that offer improved performance, and that are designed to be recycled or reused instead of landfilled where further impacts (air, water, soil pollution, and the emission of methane gas) can occur. Sustainability in commerce includes integrating such considerations into the design, sourcing, manufacturing, and distribution processes along with practices that reduce negative impacts, enhance positive social, environmental, and economic impacts, or both.

Systemic examination throughout the entire process of commerce reveals opportunities to make more with less. The systems perspective requires that consequences and impacts are determined at each phase— including environmental (air, water, land, and species impacts), social (worker, community, and stakeholder interests), and economic. Tools such as industry process maps have been developed. Simple systems maps can help guide the management team to analyze all impacts, both positive and negative, for each step in the development and delivery of a consumable product or service. Process and system mapping can be accomplished in-house or developed through consultants.

CHAPTER 5

Why Industry Is Crucial to Sustainability

Previous chapters introduced the notion of sustainability, its intersection with the paradigm of systems thinking, and demonstrated how systems thinking allows for useful feedback and critical indicators to manage economic, social, and environmental performance. This chapter examines the human groups making decisions that have the power to restore or continue to degrade larger systems. Businesses are not the only player, and individuals, government, and other organizations have roles to fill, yet the commercial sector is best positioned to both contribute to and profit from its attention to sustainability.

The chapter begins by looking at the roles that individuals, NGOs, governments, and industry play, for better or worse, while building the case for thinking systemically at each level. In the sections below some of the opportunities, barriers, examples, and best practices for each of these groupings are outlined, followed by an examination of the primacy of the industrial sector as best positioned to generate quantum leap progress is crucial to sustainability. A number of reasons will be offered as to why the participation of commerce in humankind's pursuit of a stable earth is both necessary and inevitable.

Levels of Impact

Individuals

With 7 billion people on earth making daily decisions that have consequences for local and global ecological conditions, and for economic and social welfare, many small decisions accumulate into huge ramifications. The decisions of those in the developed nations have a larger impact than those in the developing world, because the former use far greater resources than the latter.[1] The United States, for example, while populated by less than 5% of the global population, uses 32% of the world supply of corn, 24% of oil production, 23% of electricity production, 22% of natural gas production, and 16% of coal production.[2]

Global population growth is one reason for the ever increasing demand on the biosphere's resources. A growing worldwide culture of consumption has taken a toll on the planet. Someone who had turned 40 in 1972 and lived to see their 80th birthday in 2012 would have lived in a time in which 5 billion new people were born into the world, competing for the same finite set of resources. Human decision making, even at the individual level, is complicated and the science aimed at understanding those decisions is still developing, but is the strongest leverage point available for creating change.

In the 21st century, a wide range of indicators have begun to persuade individuals that they are one of a growing many in a diminishing planet of not enough. Global Footprint Network reports that as of 2008 humanity's total ecological footprint was estimated at 1.5 planet earths, meaning that humanity was using the earth's renewable ecological systems 50% faster than their regeneration rates. The report further predicts the rate to accelerate to 200% faster than regeneration rates by 2050 if business as usual continues on its current path.[3] The decisions each person makes are important; if enough people take steps to reduce their energy and resource use, civilization can begin to sufficiently address the damage humankind is causing on the planet.

As individuals, people's decisions are influenced by community and cultural values, government policy, and enterprise solutions and marketing. Also, as important as these daily individual decisions are, the impact of any individual decision is minute in comparison to decisions made on behalf of larger social groups.

The Non-Profit Sector

Representing a variety of the social, economic, and environmental concerns of society are NGOs. Operating from the tiny local to the transnational global scale, these NGOs influence public, government, and corporate bodies. Typically funded by private grant dollars from corporations, foundations, or trusts with support from individuals, the role of these organizations is to aggregate the voices or interests of their funders and to develop solutions, programs, or offerings to satisfy those interests. They serve as a unified voice for their members, generators, and translators of research, and provokers of government and corporate policy. These are groups that raise warning flags, document problems, and offer or propose solutions. At times, they create conflict with segments of society, with governments and their agencies, and with industry, yet also often act as consultants, advisors, or mediators.

As a collective of individuals all seeking the same goals, by their nature NGOs operate at a larger scale than individuals. There are many NGOs whose size and scope are substantial enough to exert pressure, either through member activism or financial resources. For example, organizations such as the Climate Action Network provide support and resources to a worldwide network of over 700 other NGOs in more than 90 countries, working to promote government and individual action to limit human-induced climate change to ecologically sustainable levels.[4]

NGOs are a social structure designed to complement the efforts of government, industry, and individuals; they are a critical piece of the system, filling gaps not addressed by other sectors. This segment of human effort adds layers of complexity to the sustainability challenge beyond that of individuals. Understanding how an NGO's needs and motivations drive their interaction with each segment of society can lead to minimizing the obstacles and maximizing their potential contribution to assisting with sustainability concerns.

NGOs can at times have trouble playing nicely with one another, as they compete for funds from donors, trusts, foundations, and individuals. For those in private enterprises partnering with NGOs around the world, it is useful to understand that there is considerable variation in how they operate. Operating overhead can consume a large share of corporate donations.

Even so, there is real substance within select programs that have organized and coordinated unified voices in the sustainability conversation. For example the World Wildlife Fund's Earth Hour Initiative has grown to more than 7,000 cities and towns from 152 countries and territories in 2012, each of which turned off their lights on March 31st, from 8:30 pm to 9:30 pm.[5]

Bill McKibben's 350.org is an organization whose mission is to raise awareness of anthropogenic (man-made) climate change. The organization established the International Day of Climate Action during which over 5,200 synchronized demonstrations in 181 countries took place on October 24th, 2009.[6] The Carbon Disclosure Project has been successful in working with 3,000 of the largest corporations in the world to ensure carbon emission reduction strategies are assessed at the operational level.[7] NGOs have a critical role in addressing sustainability at the very highest level by informing their members and the public, providing programs, and as a critical partner with the for-profit sector, as well as governments and their agencies.

Government

Most national governments are representative of the collective interests of citizens from all corners of their countries, with those serving in elected positions entrusted to create or amend legislation that serves their electorates. Whether issues are social, environmental, or economic, there is almost always disagreement among those elected and appointed representatives. All of them want the same general ends: security, prosperity, justice, and freedom. But there is little consensus about the best way to achieve these ends, and there are tensions within underlying beliefs about the role of government, as well.

For example, few would argue with the notion of increasing financial wealth (although there is a growing economic "degrowth" movement), but the means to achieve it are debated. Is the best route through increasing job training, or mandating a federally controlled savings for retirement, or decreasing regulations on industry, or directly stimulating business investment, or bailing out businesses at risk of failure? Which policy best serves national health, a health savings account, or a federal healthcare policy, or more tightly regulating the food, health insurance, or medical industry, or some other means? Policies and programs to support social well-being have been the arena for some of the most divisive legislative battlegrounds.

Of course, politically, the topic of the environment has been hotly debated, too. As with other issues, party lines are drawn. Does environmental protection take shape as regulation, which can become a cost to

business, or does it stimulate business to innovate? Environmental degradation at the hands of industry has mostly incurred little to no penalty for
the businesses that have engaged in it, yet it has exacted a cost to society
and governments in the form of extensive cleanup campaigns, escalated
healthcare costs, squandered resources, and lost habitats, not to mention
the suffering of those impacted by these practices.

As a political subject, in the United States in particular, climate change
is a sharply contested issue. Climate science is challenged by some for
what is perceived as a lack of trend data and legitimacy. There are problems related to terminology and genesis as well: is it global warming or is
it climate change? Is climate change a real trend, is it man-made or a natural cycle? (The planet is, on average and by decade warming, but there
are other aspects of climate change aside from this warming trend such
as rising sea levels, ocean acidification, extreme weather events, droughts,
floods, and other impacts.[8] Scientific consensus is that it is a real trend,
and it is caused by human activity.[9])

All political debate is about priorities. Social scientists have shown
that people tend to prioritize based on immediate and short-term concerns. Typically, that has meant the focus tips toward economic policies,
especially those that might generate a faster impact, rather than attending
to a longer time horizon. Such decisions are often incurred at a cost to the
biosphere and to society. There are many means to create prosperity, but
all nations, governments, businesses, and people are dependent upon a
set of biospheric resources and services, and a stable social infrastructure.

Many proposed sustainability solutions are of a regulatory or restric-
tive nature, which is considered by many to infringe upon their freedoms
and rights. While some legislative acts, such as the Environmental Protection Act and the Clean Water Act, both in the United States, have been
indisputable successes, there have also been regulations that were ineffectual, have impeded innovation, or resulted in unintended consequences.
As bad news is almost always more memorable than good, many people
lump any type of policy or regulatory measure as more likely to fall into
the first category. All of these considerations, though, may be designed
into policy efforts that aim at increasing innovation, offering appropriate
incentive structures and limits based on a researched balance of environmental, social, and economic forecasts.

While significant progress is still forthcoming at the national level of most governments, there is a lot of action at the state and local level, often independently, but sometimes aided by national policies. At whatever level, the role of government is to act on behalf of its constituents, both individuals and industry, to help shape a secure and prosperous future. But arguably, the player with the greatest potential for creating a more sustainable world lies with those in the commercial sector.

Industry and Its Unique Position

In May 2005, General Electric's CEO Jeff Immelt announced a new corporate initiative called "Ecomagination," intended "to develop tomorrow's solutions such as solar energy, hybrid locomotives, fuel cells, lower-emission aircraft engines, lighter and stronger durable materials, efficient lighting, and water purification technology."[10] One of the first commercials to publicly advertise this initiative in 2006 was called "Singin in the Rain." The commercial features an elephant dancing through the rainforest in Gene Kelly fashion. The narrator speaks over the classic song: "Water that is more pure, jet engines, trains and power plants that run dramatically cleaner. At GE, we're using what we call Ecomagination to create technology that is right in step with nature."[11]

This new GE brand campaign signaled that a company synonymous with products that jeopardize environmental quality was now forging a new initiative to place the environment near the top of their priorities. Such a statement was bold, and was immediately challenged by critics.

The announcement prompted an op-ed piece in *The New York Times* that said "while General Electric's increased emphasis on clean technology will probably result in improved products and benefit its bottom line, Mr. Immelt's credibility as a spokesman on national environmental policy is fatally flawed because of his company's intransigence in cleaning up its own toxic legacy."[12] Regardless of the quality of the commercial and the imagination of one of the best advertising agencies in the world, it was apparent that words and images were not enough to paper over a tradition of pollution and environmental degradation, a reputation that GE may have not been fully aware of.

GE countered by promising to invest \$1.4 billion in clean technology research and development in 2008 as part of its Ecomagination initiative. In 2008, the scheme had resulted in 70 green products being brought to market, ranging from halogen lamps to biogas engines. In 2010, GE continued to raise its investment by adding \$10 billion into Ecomagination ventures over the next 5 years.[13] As proof of this investment, GE celebrated its 20,000th wind turbine installation in 2012 in conjunction with its 10-year anniversary in the wind industry. The company also reports that Ecomagination revenue continues to grow at twice the rate of total company revenues, representing \$25 billion at the end of 2012.[14]

The GE Ecomagination business strategy was designed to rebuild a brand by driving profitability through innovative environmental solutions to technological issues. The company identified areas they could improve upon and developed a storyline to complement those activities. GE harnessed its power in industry to create a competitive advantage under the umbrella of environmental affairs. Ecomagination is one example of how the commercial sector is particularly poised to tackle the topic of sustainability.

Businesses create jobs and partner with government and the non-profit sectors, while also being held accountable by all of these stakeholders. Industry is a force that develops and refines the tools and resources that have made lives longer, safer, easier, and more interesting. It is a powerful shaper of opinions, images, and appeal. And it requires a reliable, affordable set of resources to continue to thrive, as well as a population of customers to survive.

Many companies feel confident about their role in society, evident in their corporate slogans, which position them as a force for good and for progress. For example,

Dell: "Get More out of Now"
Ford: "Built for the Road Ahead"
Hitachi: "Inspire the Next"
Panasonic: "Ideas for Life"
Walmart: "Save Money, Live Better"

Commerce is both the biggest challenge to sustainability and the area with the greatest potential to create its solutions. It is the sector that has been the most significant contributor to sustainability problems, yet it is also the sector with the capital, speed, flexibility, and innovation to solve those problems. Many businesses joining the sustainability bandwagon approach their strategies from a reactive and incremental perspective, taking advantage of low-hanging fruit and its reputational advantages.

Such an approach is a good initial step, one that by necessity will need to be followed with a more radical and transformative approach that pushes institutions and the public to shift the way they think about society and the natural world. Radical solutions seek to find ways to restore or regenerate natural capital or increase the standard of living for the billions of people who live in poverty while reducing their burden on the planet. At the same time they offer immense possibility for increasing economic capital.

Other transformational arenas include the creation of new markets and of disruptive technologies, products, and services employing biomimicry, the development and further deployment of feasible, affordable energy sources, and the transition from offering product to offering service.

While industry is the sector bearing the greatest responsibility for generating the most substantive damage, the problems being faced are not solely the responsibility of industry to resolve. Businesses will need the incentives designed by governments, the guidance and assistance from knowledgeable and experienced NGOs, and the adoption and participation of the public. Enterprise is the most potent force capable of taking the lead.

Organizational scholars have recognized that the social and environmental problems facing society "cannot be considered, much less resolved, without the inclusion of business as a central factor."[15] Others studying the subject have agreed that the cooperation and involvement of business is necessary[16][17] and inevitable.[18] Here are some reasons why:

- *Speed*: Of all sectors, the for-profit sector is known for its speed. Governments and NGOs, on the other hand, are known for their slow, cumbersome, bureaucratic machinations. The rate of destabilizing change in economic, social, and environmental systems has been quickening, and industry's structure and culture is best suited to expeditiously respond.

- *Knowledge*: As technological innovators, companies best understand the economic and technical tradeoffs involved. No other sector combines the depth of financial understanding with the engineering feasibility and market viability.
- *Expertise*: Companies must be involved in regulatory and policy decisions, as government agencies do not have the knowledge or resources to develop the best solutions. While some government agencies hire from industry, policies are likely to fail or backfire if they do not incorporate the knowledge and expertise of a range of organizations in the industries for which they are designing those policies.
- *Influence*: As social structures, businesses, industries, and markets have accumulated the power and resources to influence economic, social, environmental, and political conditions, and have been involved in developing solutions to problems in these realms. Through services, products, and marketing, the commercial sector has immense sway over consumer and business-to-business behavior.
- *Incentive*: Businesses can profit through creating innovations to satisfy societal preferences for products and services that resolve social and environmental problems. Incentive to capitalize on solutions to these problems is built into commercial activity, unlike the other sectors.
- *Innovation*: Where governments are beholden to their voters, and NGOs to their membership, enterprise is the only sector with sufficient latitude to continually experiment with new technologies, products, and services. While they must answer to boards and shareholders, as well as other stakeholders, and may be punished for failure, business interests are more resilient and flexible when it comes to experimentation and innovation.
- *Result orientation*: Businesses are driven by results, with a sharp focus on strategic approaches, selecting key performance indicators, and monitoring and revising processes and practices to enhance outcomes. Governments and NGOs are not held to as rigorous an approach to performance.

- *Access to capital*: In addition to knowledge and expertise, no other sector enjoys as ready access to financial and human capital as does industry.
- *Collaboration potential*: Industry trade and peer group cooperation, inter-industry alliances, and inter-sector partnerships tie into the characteristics discussed above to enhance the potential for improvements to processes, services, products, and even business models, in the pursuit of enhanced sustainability.

Benefits of Corporate Sustainability Practices

The evidence for the business case for sustainability is clear. The specific benefits of corporate sustainability practices for any particular organization, however, can be as unique as each firm implementing sustainability. For example, Warren Buffet's Berkshire-Hathaway-owned Shaw Industries has created a facility intended to reclaim used carpets for new raw materials, reducing the need for additional raw materials. Shaw developed and manages what it calls its Evergreen facility where post-consumer Nylon 6 carpet is turned back to its original starting material, caprolactam.

Procurement of more sustainable raw materials has also led Mohawk carpets to invest in facilities that source raw materials from readily available post-consumer plastic bottles to create PET fiber. Most carpet fibers are oil based and therefore subject to cost volatility due to the fluctuating price of crude oil. Creating a process to vertically integrate and reduce cost risks in the supply chain is one benefit of sustainability.

Another benefit is using it to refocus branding so as to enhance reputation. Walmart's supply chain initiative is an example of how one company used this strategy through their influence on the company's vast supply chain. Walmart's sustainability story is often begun with how the company used sustainability as a proactive strategy to shift focus from labor issues to environmental issues. In 2006, the company made a highly visible commitment to sustainability prompting a spate of media dialogue questioning how the retail giant could position itself an environmentally and socially responsible company. In spite of the criticism, and while acknowledging that attention to the social aspect of Walmart's sustainability efforts has been comparatively weak, the company has made substantive strides in the environmental leg of the TBL stool.

Walmart conducted a Greenhouse Gas Inventory of its operations and supply chain, the outcome of which showed that the majority of the company's environmental impacts came from the nearly 60,000 suppliers who put products on the shelves of its stores. The corporation developed a program to address these supply chain issues. Their Supplier Sustainability Assessment, launched in 2008, comprises 15 scored questions that provide Walmart with a window into the sustainability programs of its suppliers. The survey questions are separated into four categories that include energy and climate, material efficiency, nature and resources, and people and community.

Reducing environmental impacts in the name of sustainability is not the only strategy Walmart has at play. Working with its suppliers, the notoriously price-slashing retailer now has a new competitive advantage with its suppliers. Rather than purely debating price with suppliers, Walmart has laid the foundation to provide its buyers with data that can be used to further slash prices. For example, if a supplier has reduced energy costs by 20%, it is probable that operating costs have also decreased.

In 2012, Natural Capital Solutions released a report summarizing over 40 studies that prove the business case for sustainability.[19] The studies covered in the report were conducted by academic institutions, conventional management consulting firms, and similar research entities such as Deloitte, Gallup, Harvard Business Review, Goldman Sachs, and the U.S. Department of Energy. The report categorized the short- and long-term profitability strategies into five areas:

- *Efficiency and risk reduction*: Natural resource, energy, and operational efficiency included conserving capital for long-term growth, hedging exposure to price volatilities, mitigating supply chain disruptions and environmental degradation, and, of course, reducing input and overhead costs.
- *Talent*: Among human resources management benefits were improved attraction, retention, and productivity of top talent, reduced training costs and absenteeism, and greater retention of corporate knowledge.

- *Revenue enhancement and expense reduction*: Financial operations were improved through lower insurance premiums, decreased borrowing costs, improved investor relations, and increased market capitalization and stock growth.
- *Branding and reputation*: In the marketing and communications arena, sustainability strategies paid off in the form of differentiated products with greater brand image and loyalty, and an expanded and more secure base of customers who shop more frequently.
- *Stakeholder relations improvement*: Benefits to firms in partnering with government and NGOs include strengthened regulator and community relations, reduced risk to brand reputation and of government sanction, access to new markets, and enhanced supply chain management.

The advantages of integrating sustainability practices are not the sole province of big business. Small and medium-sized enterprises have seen similar advantages. For example, Custom Packaging, a Lebanon Tennessee-based packaging company, has adopted aggressive sustainability practices. Their projects have included the largest solar array in their county, a zero waste manufacturing policy, and energy conservation practices that have resulted in energy intensity per square foot that is lower than most Leadership in Energy and Environmental Design (LEED) certified buildings. Beyond implementation of these programs, the family-owned business is exploring alternative packing materials such as mushroom, bulrush, and bamboo to offer as replacement to traditional corrugate.

In Vancouver Canada, Rocky Mountain Flatbread, a pizza restaurant, wholesale business, and mobile caterer, has been acting locally to drive sustainability. Ninety percent of the eatery's produce is locally sourced including food grown by local schools at which the restaurant is involved in teaching students food cultivation. Rocky Mountain Flatbread also has a goal to create a zero waste menu and to provide carbon neutral food. The company limits its emissions and buys carbon credits to offset the balance. Cofounder Suzanna Fielden believes "sustainability and profitability are parts of the same approach."[20]

UKOS, a UK-based stationery and business supplies provider, has not only reduced energy costs by an average of 5% year-over-year for the last 4 years, but has also increased sales and profits by differentiating itself as a sustainable supplier. UKOS was at the top of the "The Sunday Times" Best Green Companies List in 2010.[21]

Musco Family Olive Company, the largest retail olive producer in America, has developed a closed-loop waste program called RENEWS (Renewable Energy and Wastewater System). In producing canned olives, Musco amasses millions of olive pits. RENEWS cleanly burns 15 tons of pit waste daily, using the heat to evaporate spent process water and create steam. The steam drives the largest production steam engine in the United States, which then powers the plant. Every year the system uses 13 billion olive pits and helps Musco recycle a significant amount of its water.[22]

In their 2011 Corporate Sustainability Report, KPMG Cooperative International said sustainability "is a powerful undercurrent running through the pages of the business media, an almost compulsory topic of discussion at meetings of business leaders, and among the most thoroughly researched business issues of the past decade."[23] Executives and managers who ignore the trend do so at their own, and their firm's, growing disadvantage. Those who view the trend as a burden and inconvenience are likely to have weak performance and weaker benefits.

Sustainability *leaders*, companies that have learned to harness sustainability as a viable business strategy for their organization, recognize that there are strong benefits to integrating sustainability practices deep into their operations and business models. Sustainability *followers* have gone the incremental, add-on-program route in the process of implementing strategies that focus on TBL principles. Sustainability *laggards* who have failed to adopt the principles of sustainability altogether are exposing themselves to risk, but trends are in the process of creating increasing pressure from the public, governments, NGOs, and the media.

The Future of Corporate Sustainability

A vision for the future of sustainability can inspire innovation and positive disruption. This section covers more recent developments and initiatives in the sustainability space.

"Beyond Compliance" is a term in the sustainability arena used to characterize firms that go further than mere legal compliance with regulations or expectations, to be more thorough, deeper, and systemic in the application of their practices. It's a proactive strategy to address potential costly future regulation, market erosion, resource constraints, customer and employee loyalty, insurance requirements, and investor concerns.

One new movement that encompasses such a mindset is the "B Corp" or benefit corporation. Benefit corporations work social, environmental, governance, and economic considerations into their for-profit organizational charter. They are certified by the non-profit B Lab (www.bcorporation.net) as meeting rigorous standards of social and environmental performance, accountability, and transparency. As of the end of the first quarter of 2013, 12 U.S. states had passed legislation providing for benefit corporation charters, and 20 states had legislation pending.

Globally, there are 600 Certified B Corps from 15 countries and 60 industries. Together, they are working to redefine the success in business, for example, from "best in the world" to "best for the world." The benefit corporation movement positions itself as a concrete, market-based, and scalable solution to systemic challenges.

Another trend is the hybrid organization, in which sector partnership or crossover exists between, for example, a government body or agency and a for-profit business. Another version entails an NGO incorporating a set of for-profit components to their revenue stream, and permutations of sector-crossing innovation. Such organizations embody a values-driven or mission-compelled ethic and they are creating a shift in markets, blurring previously distinct sector boundaries. They do this by creating innovative ideas about supply chains and sources of capital, as well as entire new business models, using capitalism to solve the world's social and environmental problems. These are not fly-by-night or short-lived organizations, as detailed in books such as *Hybrid Organizations*[24] and the *Oikos Case Collection*.[25]

A similar concept is social or sustainable enterprise, a bigger umbrella that includes both the benefit and hybrid organization, but goes beyond these models. Such an enterprise might be a dedicated division of a for-profit corporation. One such example is Procter and Gamble's PuR water filtration product line marketed and distributed through a private sector approach to community-based outreach.

Sustainable enterprises might instead be a small local business, or just about any size, scale, or sector format or, for that matter, combination of sector formats. To qualify as a sustainable enterprise, the mission and model must directly address environmental and social problems, locally or globally, must address TBL practices in their operations, and must have a viable and sustainable economic model, even if that model is from a combination of sales, grants, contributions, and other income.

The innovations detailed here are growing because of a widening recognition that delay on the issues they address is going to be expensive, both for businesses that eschew the sustainability trend, and for global society. As any procrastinator has experienced, the longer something is put off, the deeper the hole becomes and the more effort, expense, and struggle are required to climb out.

The next chapter explores the mechanisms that firms use to determine sustainability strategies, both data-driven and more subtle strategies in the commercial sector's efforts at using sustainability as a tool for driving business performance.

Chapter Summary: Key Takeaways

Industrialization, made possible by scientific discovery and the use of fossil fuels, made rapid population growth possible, stimulating further industrialization and resource use. Slowly, beginning a century and a half ago, and much more rapidly in the last four decades, has come a realization that industrialization, population, and technology are destabilizing forces on the biosphere.

Responsibility to restore the planet's dynamic equilibrium is shared among 7 billion individual humans (with those in emerging nations using a fraction of the resources of those in advanced nations), non-profits, government, and for-profit business. The combined efforts and partnerships between each of these sets of entities will be required.

Industry, however, is both the biggest driver of destabilization and the most capable player in taking the lead. It is the sector with the speed, knowledge, expertise, influence, incentive, innovation, result orientation, access to capital, and collaboration potential flexibility, and innovation to solve the big problems facing humankind.

The benefits to industry for doing so, though, are many, including conservation of and better access to capital, long-term growth, expense reduction, employee engagement, brand recognition, and customer loyalty, to name a few.

With the advent of hybrid and B corporations, enterprises may be adjusting to the need to balance profit primacy with environmental health and social well-being.

CHAPTER 6

Corporate Sustainability Frameworks, Metrics, and Indices

Earlier chapters were designed to provide a working understanding of sustainability and what it means for business, why it is so important, how to think systemically about business processes, the risks for ignoring sustainability trends, and the rewards for addressing them. This chapter provides an overview of some commonly used metrics and tools, while pointing out the need for customization of any plan to incorporate each individual company's industry, size, mission, vision, and business model.

Nobel Prize winning author George Bernard Shaw wrote in, *Man and Superman* in 1903 that *"the only man I know who behaves sensibly is my tailor; he takes my measurements anew each time he sees me. The rest go on with their old measurements and expect me to fit them."*[1] In practice, sustainability resembles the role of a tailor, as it requires regular measurement, modification, and adaptation. Organizations grow in size at some times, contract at others, change shape, and improve in health. The sustainability function, like the tailor function requires a keen attention to detail.

As a custom suit that must be tailored for the client, sustainability initiatives must be designed to fit the needs of each company. What works for Ford may not work for GM, Pepsi may be different from Coke, the approach of Ernst and Young should and could be different than that of Deloitte. Varying needs require different approaches and tools to be successful. Where the tailor's tools include measuring tape, scissors, pins, and chalk, a sustainability practitioner has a number of tools as well, including operational and reporting frameworks, GHG inventories, LCAs, carbon disclosures, and other tools. These resources and approaches address planning, information gathering, precision, and transparency. Some of them establish the baseline for comparing the degree of progress or failure over time. When compiled and used properly the data prompts firms to weigh the consequences of differing options, from both quantitative and qualitative approaches, to find the most feasible solutions.

Decision making employing analytics and materiality ensures that environmental and social impacts are measured and monitored over time, showing clear and proven trends and accomplishments. Without such considerations, programs portrayed as sustainable are likely to miss the mark by being simply ad hoc or feel-good efforts. A delineated and comparative set of criteria based on a systemic review of inputs, outputs, and processes, makes the difference between a flawed and ineffectual decision and one that has much greater sustainable value.

Evaluating the degree of sustainability for an organization, product, or service is a fairly complex task. One example is how the embodied energy of 1 kWh can differ by state: there is a difference between the carbon footprint of 1 kWh generated in Colorado and 1 kWh generated in Tennessee. Another example is how a "sustainable" product may not perform exactly to specification depending on use. Length of use determines the sustainability profile of carpet tile versus broadloom. A live-cut Christmas tree may appear "greener" than an artificial tree, but an LCA may prove otherwise. Sustainability metrics change by zip code, function, and use. These and similar questions make it necessary to employ tools to collect data and generate analytics, in turn establishing a foundation for sustainable decision making.

The tools and frameworks discussed in this chapter are some of the most frequently used by firms pursuing more sustainable practices. All are science-based. Each is introduced very briefly here and discussed in more detail in the sections that follow:

- The Natural Step Framework is a comprehensive model for sustainability planning in complex organizational systems, based on four central principles, called system conditions, and the backcasting method of planning.
- The Circular Economy is a framework for redesigning business practices toward an efficient and restorative model.

- Greenhouse gases are measured though a greenhouse gas inventory, also called carbon accounting.
- Lifecycle assessment is a scientific method of measuring the impact of a product or service across all stages of the product lifecycle from raw materials through end of life (known as cradle to grave).
- Carbon disclosure frameworks offer firms another means to assess environmental risks. Responding to these tools is an exercise that requires a hard look at climate, water, supply chain, and forest vulnerability and opportunity for any organization.
- The Dow Jones Sustainability Indices evaluate the financial, environmental, and social performance of the largest 2500 firms listed by the Dow Jones Global Total Stock Market Index, and are tools for investors and the financial industry.
- The GRI provides another option for a voluntary wide-scope framework for organizations to follow when creating a sustainability report encompassing economic, governance, social, and environmental indicators. More than 4,000 organizations from 60 countries use the GRI guidelines to produce their sustainability reports.
- Consumer-oriented indices and certifications such as Walmart's Sustainability Index, the Higg Index created by the Sustainable Apparel Coalition, Green Seal certification, and the EcoLogo program are other options.

The Natural Step

The Natural Step is a non-profit organization that helps organizations with their sustainability training and planning, using their Framework for Strategic Sustainable Development. Its popularity is partly due to its simplicity, as it distills sustainable practices down to four fairly short principles which are positioned as nonconfrontational, nonprescriptive, and nonjudgmental. These principles act as criteria for developing a sustainable society and, by extension, a sustainable set of business practices. They are worded as four systems conditions:

"In a sustainable society, nature is not subject to systematically increasing:

1. concentrations of substances extracted from the earth's crust,
2. concentrations of substances produced by society,
3. degradation by physical means,
4. and, in that society, people are not subject to conditions that systemically undermine their capacity to meet their needs."[2]

This doesn't mean an immediate and permanent cessation of mining, producing GHGs and toxic waste, otherwise damaging natural systems, and addressing every incidence of social injustice. It serves as a template to immediately begin to find ways to reduce and eventually eliminate these threats. In practice, this framework focuses organizations on finding alternatives to the use of heavy metals, fossil fuels, toxic chemicals, and pollutants, over-harvesting resources, the reduction of biodiversity, and a variety of social impacts such as unfair and unsafe working conditions, dangerous products, corruption, and anti-competitive behavior.

There are two rationales underlying the system conditions. The first of which is that most harmful materials extracted or produced do not disappear, they simply disperse into the environment, often accumulating to rates that cause widespread long-term harm. The second rationale is that, as part of a larger system, business practices must operate in alignment with the cyclical characteristic of natural systems so as to avoid destabilization of these systems and cycles. Attention is paid to avoiding the generation of new problems in the process of solving existing ones.

To achieve these systems conditions, the framework employs a planning approach, called backcasting, which begins with the end in mind, a vision of the desired condition at a future point. Using that vision and time frame, milestones, with strategies, action steps, and monitoring procedures, are projected backward from the end point to the present. This creates a template to move step by step toward the desired vision in which each step is associated with its own payoff.

Once a vision is determined and targets are set, current practices are evaluated to establish a baseline to serve as a starting point for a gap analysis of services, products, human and financial capital, and energy flows. The framework encourages creative solutions that will lead to more radical, versus incremental, change, through its emphasis on the guiding longer-term vision.

The Natural Step is one way of conceptualizing a systemic understanding of how business practices fold into a larger global world. Greenhouse gas accounting, covered below, through its implicit tie to climate change, is also systemic in nature, although perhaps less obviously and comprehensively so.

The Circular Economy Framework

Based on a systems-oriented cradle-to-cradle perspective, the Circular Economy is a framework developed by the Ellen MacArthur Foundation using a number of schools of thought which share the same basic principles. With the purpose of creating a restorative model of industrial process design, biological materials are returned safely to the biosphere and technical materials are prevented from entering it, as espoused by waste experts McDonough and Braungart. Biological materials are consumed, versus technical materials, which are used, then recycled and reused or repurposed, with all materials managed in a cyclic flow.

The Circular Economy framework principles are founded on a living systems model and include the use of only renewable energy resources, building resilience through increasing or prioritizing diversity, the elimination of any waste that is not part of any natural biological cycle, and the cascading of materials from one application into a next use.[3] These last two principles, while eliminating any toxics, envision biological outputs, which are deemed as waste by traditional business models, into nutrients for living systems or incorporated into production processes. For example, agricultural waste might be directly composted, or used for another purpose or product, such as livestock feed. The manure generated by the livestock can then be used as a soil amendment for regenerating degraded habitats or for the next cycle of agricultural products.

Greenhouse Gas Inventory

Greenhouse gas (GHG) emissions are atmospheric gases, mostly generated through the burning of fossil fuels such as coal, oil, and natural gas, which absorb and emit heat. The primary sources of these GHGs are through electricity generation, transportation, industrial processes, commercial and residential energy use, and agricultural production. GHGs trap heat, fueling the greenhouse effect, the process responsible for the overall global warming trend. The last 150 years of increasing industrial activity correspondingly elevated the concentration of these atmospheric GHGs.

At the dawn of the Industrial Revolution in the mid-1700s, the world's human population grew by about 57% to 700 million, and reached 1 billion in 1800. The birth of the Industrial Revolution altered agricultural technology, and ushered in medical improvements and a rise in living standards, resulting in the population explosion that steamrolled into the next centuries. Within 100 years of the onset of the Industrial Revolution, the world population grew 100%, to 2 billion people in 1927. During the 20th century, the world population would continue to expand exponentially, growing to 6 billion people just before the start of the 21st century, representing an incredible 400% population increase in a single century. In the 250 years spanning the Industrial Revolution, the world human population has increased by 6 billion people.[4]

The explosive growth of humanity is pulling the planet much closer to the boundaries of its human-supportive dynamic equilibrium. In the scientific literature, there is a strong consensus that global surface temperatures have increased in recent decades and that the trend is caused primarily by the emissions of GHGs emitted by people. Prior to the 21st century, GHGs were mostly generated by the developed nations, but now developing countries are catching up and even surpassing them. Disputes over the key scientific facts of global warming are now more prevalent in the popular media than in the scientific literature, where such issues are treated as resolved.[5]

Despite that, some still question whether population-induced climate change is, in fact, happening there is still benefit in the use of GHG inventorying as a tool for evaluating corporate performance and (one facet of) environmental impact. As an indicator, the measurement of GHGs provides a method to analyze behaviors, benchmark the footprint of multiple facilities against one another, or determine which environmental issues are most material. Conducting a GHG inventory helps quantify an organization's overall contribution to climate change, and build models in which rates of climate change, and activities to mitigate those impacts, can be evaluated. It also identifies opportunities to reduce inefficiencies, and thereby reduce costs, and points out environmental impacts.

Sustainability consultants can provide assistance with GHG inventorying. For firms who desire to pursue it in-house or seek more information, guidance for inventorying GHG emissions on the organizational level can be found through the World Resources Institute and World Business Council for Sustainable Development GHG Protocol. For national GHG inventories, guidance is provided by the Intergovernmental Panel on Climate Change methodology reports.[6] For local governments, The Local Government Operations Protocol is a tool for accounting and reporting GHG emissions across a local government's operations.

Greenhouse gas analytics offer an indicator that, if measured and reported properly, can be as useful as financial accounting data. Processes can be evaluated based on energy consumption, water usage, waste generation, and even purchasing data. The purpose and rationale behind the employment of GHG accounting tools determines their usefulness. The following example shows the difference between two organizations that were completing GHG inventories. Both companies are in the textile industry and produce complementary products.

Company A looked at inventorying their emissions as a chore, a check-the-box endeavor that was a requirement for a product certification that the company was seeking to obtain. It collected data at the campus level for each of its five campuses, and included water, electricity, natural gas, fleet, and waste data for each facility. Lack of foresight and care produced errors in the collection and input of the data, including transposed information, extra keystrokes, missing information, and inconsistent units of measurement. The company successfully completed an inventory that met the requirements of the World Resources Institute GHG Protocol but couldn't answer the simple question, "Which of the five facilities had the biggest footprint?"

Company B's senior leadership, in contrast, wanted the inventory done right the first time. There were no existing systems through which utility data was captured, most data was decentralized and had to be collected from individual managers. The data capture encompassed four facilities, all within the same city limits. Each facility had one set of meters but, thinking with the end in mind, company B collected data at the process level as well.

The inventory of both companies met the requirements of the protocol recommended by the World Resources Institute and was considered permissible in a voluntary reporting world. Only one was useful, though, when trying to determine emission-reducing activities and monitor emissions over time, or in prioritizing limited resources. The only difference between Company A and Company B was how data was segmented and then re-aggregated. The design of a GHG inventory process is the most critical step in creating useful sustainability analytics for decision makers, and will be discussed in more detail in Chapter 7.

At the industry trade level, the Automobile Industry Action Group, a non-profit organization, works to be *"a catalyst for the global automotive industries efforts to establish a seamless efficient and responsible supply chain."* Within the trade group's Corporate Responsibility Committee is a subcommittee working on GHG inventorying and reporting issues. Led by co-chairs from GM and Ford, this group has designed a mechanism to develop a common GHG emission calculating and reporting process. With the decline of vertically integrated automakers, traditional automobiles are made using a network of suppliers making specific original equipment manufacturer (OEM) parts for each vehicle make and model. Often more than one supplier makes a part for each make and model. The trade group reporting mechanism provides emission data, which allows a comparative evaluation to determine which suppliers are making products that generate fewer emissions.

While a GHG inventory offers a comprehensive tool for evaluating some environmental aspects of sustainability, it is not a panacea nor is it quick or easy. The process can be expensive and time consuming. To manage vast amounts of data, organizations may opt to purchase carbon accounting software, although this may add to expense and delay in implementation. Some providers offer carbon accounting software to provide more affordable options with less functionality. For smaller organizations, this effort may simply not be worth the time. Single facility companies, smaller service companies, and firms that lease their facilities and pay only electricity bills may be better served by focusing on other aspects of sustainability. For larger organizations with multiple sites, facilities and processes, a GHG inventory practice can be used much more effectively as a tool for assessing impact.

Lifecycle Assessments and Other Environmental Metrics

The operations of all organizations have both vertical and horizontal activities. A GHG inventory considers environmental impacts of an organization's vertical activities, its operations, and all activities required to run the organization. For almost every organization, especially those in manufacturing, that is only a partial assessment of their environmental impact. To capture the full measure, the horizontal impacts must also be evaluated.

The LCA evaluates the impacts at all stages of a product's lifecycle from cradle to grave, from the sourcing and extraction of the raw materials to the product's end of life, destruction, reuse, or decomposition. The idea behind this assessment is to capture and identify the most material sources of impacts across that horizontal lifecycle. Lifecycle assessments are a tool for credibly defining and evaluating environmental claims based on sound, peer reviewed scientific data.

Recall a consideration from earlier in this chapter: which has the smaller environmental footprint, a 6.5 foot tall live-cut Douglas fir Christmas tree grown in North Carolina or a 6.5 foot tall artificial Christmas tree sourced from China and made from PVC? At first glance, there may be little doubt that the live-cut tree is the "greener" option as it is (or was) a living organism. Prior to its harvest, the live-cut tree was also capable of cleaning air, stabilizing soils, and sequestering carbon. Upon disposal the tree can be turned into mulch, the foundation of a sand dune, a habitat for fish in lakes, or even as fuel for a fire. The live tree is also likely harvested within a few hundred miles of its purchase and use, in comparison to the transportation impacts of the artificial tree from China to the United States.

It would appear that the live tree is the obvious winner, and on a one-to-one basis the live-cut tree is sustainably superior. It is the more environmentally friendly option when comparing one of each tree. The use of the LCA enables firms to measure the degree of "greenness" of each tree more precisely. The assessment quantitatively analyzes indicators such as primary energy demand, embodied energy, acidification, eutrophication, global warming potential, and contribution to smog.

Within the LCA other variables can be added, such as the distance the seed has to travel from the production area to a nursery, from the seedbed to transplant bed, the nursery to plantation, plantation to farm, and farm to retailer can be considered. Another component of the cradle-to-grave journey is the consumer's round trip to purchase and bring home the live-cut tree. Once disposed of, end of life scenarios must also be accounted for whether the tree was composted, incinerated, or sent to landfill.

The same level of analysis for the artificial tree begins with incorporating the impact of the sourcing, refinement, and assembly of wires, PVC resins, injection molding processes, steel poles and fasteners, tape, and cardboard boxes. The transport of these parts tracks them to the factory, from the factory to the harbor, from their country of manufacture to their country of use, from a harbor to storage or a distribution center, from there to a retailer, and from a retailer to a home.

The level of detail in an LCA is valuable for anyone in marketing, product design and development, manufacturing, advertising, corporate communications, and sales. The International Organization for Standardization has created the ISO 14020 standard, which establishes nine guiding principles that a firm can use to enhance their labeling schemes and all environmental claims.

Similarly, the Federal Trade Commission offers guidance on the communication of green claims, such as the purchase of carbon offsets, use of solar power, recycling or zero waste statements, and any other green marketing claims, found in their "Green Guides." These guiding documents outline the steps required to make credible green claims, which often require LCA to be considered credible.

The Christmas tree example is from an actual study conducted in November 2010 for the American Christmas Tree Association of West Hollywood, California.[8] The study was the first ISO compliant LCA conducted with primary data, describing the manufacturing processes of artificial trees and the impacts of live-cut trees. The research was peer reviewed by academics and industry experts from Carnegie Mellon, the American Chemistry Council, and North Carolina State University.

While an LCA on a one-to-one basis shows the live-cut tree to be more sustainable, the above comparative review fails to consider one factor: an artificial tree is typically used for more than one Christmas. The results of the assessment showed that one artificial tree used for 6 years requires less energy to produce than six live-cut trees. On its sixth Christmas, an artificial tree has achieved an environmental break-even point equal to six live-cut trees. If it is used for more than 6 years, the artificial tree may well be the greener option. Folded into consideration is that some artificial tree firms offer an extended manufacturer warranty, and that the majority of artificial trees are donated, not disposed of, when first owners discard them.

Lifecycle assessments were, for a time, a less functional universal tool than one that was unique, proprietary, misunderstood, and sometimes misused. In 1991, concerns over companies using LCA results to make marketing claims prompted 11 State Attorneys General to denounce the use of LCA results to promote products. They required that such claims could not be made until a set of uniform methods for conducting such assessments was developed and consensus reached for how this type of environmental comparison can be advertised non-deceptively. Lifecycle assessment standards weren't developed until 1997, with the completion of ISO 14040 Lifecycle Assessment Standards in 2002.[9]

Even with such standards, there continues to be criticism of how LCAs are conducted and how results are used. They require a significant investment of time and resources. Data must be collected across the value chain and from suppliers who are sometimes unwilling to provide data. When this happens, data sets on specific materials or processes can be purchased from organizations such as Ecoinvent or GaBi. Free LCA data also exist and can be downloaded from websites like OpenLCA.org. Be aware that there are concerns over the accuracy of this type of acquired data set, such as how relevant the data set might be when compared to what would have been collected directly from a firm's supplier.

Another concern is the way in which organizations compare LCAs. For instance, one company might compare the results of its assessment against another company. The first might determine that the second left out a key process invalidating the latter's assessment, a result of error, or the use of different data sets like GaBi or Ecoinvent, or differing processes between companies.

Throughout the development of LCAs, the availability of data and the standardization of methods have been thorny, but standards have continued to evolve through the creation of Product Category Rules. These Product Category Rules are used to ensure that industry-specific categories have defined rules and methods for how LCAs must be conducted by companies making products for each industry. Carpet, cement, bakery products, sauces, and sparkling wine are some of the industries for which Product Category Rules have been developed. Their purpose is to limit the misunderstanding of LCAs and to increase accuracy and transparency in interpretation and in marketing claims.

Environmental Product Declarations were created to help communicate summarized LCA data in a standardized format for quantifying and describing the environmental impact of a product or raw material, used for the purposes of disclosure.[10] Declarations include information on the environmental impact of raw material acquisition, energy use and efficiency, content of materials and chemical substances, emissions to air, soil, and water, and waste generation. Product and company information is also included.[11]

These rules and declaration guidelines are gaining traction in the process of the standardization of the LCA as a tool for companies making purchasing or design decisions. Lifecycle assessment is best used by companies providing raw materials.

Service companies might also use LCA as a way to measure the environmental impacts of different service stages. Companies providing products for the built environment, such as carpet, fixtures, or ceiling tiles should consider using LCA and some of the other tools described above, given the advancements in building standards, specifically LEED (Version 4). If your organization doesn't produce a product or offer a service but purchases items in bulk, consider learning how to interpret the results of LCAs to drive more sustainable purchases.

Socially focused organizations may want to consider the use of social LCAs to evaluate social and socioeconomic impacts, or potential impacts. Guidelines, written and published by the United Nations Environment Program (UNEP), consider the impacts of production and consumption on workers, local communities, consumers, societies, and all value chain actors.[12]

Carbon Disclosure

Lifecycle assessments and GHG inventories are tools used to evaluate environmental impacts across a firm's vertical and horizontal activities to provide a solid understanding of a product's most significant material environmental impacts. This allows the firm to identify the product's "hot spots," places in their production stream that have the greatest impacts, and on which the firm is able to influence change. This also affords the opportunity to develop strategies to effectively reduce or eliminate these hot spots.

Embedded within the results of a GHG inventory and LCA are the data needed to evaluate the environmental risks of a product. CDP, formerly known as the Carbon Disclosure Project, has become an influential player in measuring, managing, disclosing, and sharing an organization's risks related to the environment, particularly climate, water, and forest-related issues. In 2010, the organization was called "the most powerful green NGO you've never heard of" on the Harvard Business Review Blog Network.[13]

The non-profit organization works with 3,000 of the largest corporations in the world to help them ensure that an effective carbon emission reduction strategy is made integral to their business. This effort is taken seriously because of the size of the shareholdings backing CDP—655 institutional investors with $78 trillion under management. They also work with cities and other government bodies.

CDP has a history of advancing sustainability programs at companies large and small, and has worked with corporations including Walmart, Tesco, Cadbury Schweppes, Procter and Gamble, and Ford to measure emissions through the supply chain. On an annual basis, CDP sends requests to the suppliers of these and other companies to complete one of their annual questionnaires, the results of which are then scored. Final scores are then reported to the requesting entity.

CDP also acts as a repository of carbon, climate, water, and forestry risk data for the world's largest economic influencers. Within this repository is information institutional investors can review to assess the climate-related risks and opportunities of organizations. This includes an evaluation of corporate risk from future legislation, environmentally focused lawsuits, or shifts in consumer's perceptions related to an organization's environmental performance. Much of this information is not collected or made available to shareholders in any other form, although in 2010, the U.S. Securities and Exchange Commission and the Canadian Securities Administration issued a Guidance Document related to climate change disclosures in annual and quarterly reports.

The Carbon Disclosure Report comprises five levels of reporting. The Investor CDP, the organization's original program, includes *"the largest collaboration of investors in the world and serves to place relevant climate change information at the heart of its institutional investors to help drive financial decision makers to migrate to low carbon economies."*[14] The Investor program handles the collection and exchange of information from a company to the institutional investor and allows investors to tie economic forces and environmental performance together to evaluate long-term performance. The Investor program is also a source for transparency and accountability, allowing investors to hold a mirror up to an organization's stated environmental commitments and goals.

Another CDP offering is its Public Procurement Program, which enables local and national governments to evaluate climate change impacts in the supply chain. The program was developed with a focus similar to the Investor program by providing essential climate data to purchasing decision makers of cities and governments.[15] In the United States, CDP is working to support Executive Order 13514. Titled *Federal Leadership in Environmental, Energy, and Economic Performance*, and signed by President Barack Obama in 2009, the order asks the General Services Administration "to investigate the impacts of requiring vendors and contractors to register with a voluntary registry or organization for reporting greenhouse gas emissions."[16] The aim of the project is to assess the cost and benefit of federal suppliers measuring and disclosing climate change data and to help local governments work toward building a low carbon government supply chain.

Building on the interest in capturing carbon data from the supply chain, CDP developed a Supply Chain Program. As of early 2013, CDP was working with 57 global corporations to encourage suppliers to disclose climate change information.[17] Companies like Walmart,

Dell, L'Oreal, and Coca-Cola are involved in rolling out the program to their suppliers. Walmart, for instance, includes a question in its Sustainability Supplier Assessment Questionnaire, asking suppliers "Have you opted to report your greenhouse gas emissions and climate change strategy to the Carbon Disclosure Project?"[18] In Walmart's case, 80% of the company's emissions originate in the supply chain of its roughly 70,000 suppliers providing products to Walmart's shelves. To influence their suppliers' practices, in 2010 Walmart created a goal to eliminate 20 million metric tons of GHG emissions from its global supply chain by 2015.

CDP for Cities is a program for local governments to report and disclose climate data. Launched in late 2011, this initiative provides a standardized reporting framework for emission data and the analysis of climate risks, opportunities, and adaptation plans for cities around the world. Initially rolled out to the C40 cities, the program invited 292 cities around the world to participate in 2013. The value of the Cities program lies in its crowd-sourcing potential. City sustainability managers will be able to identify colleagues who are addressing similar risks and issues with new and innovative strategies for reducing carbon emissions and for mitigating and adapting to risk from climate change.[19]

CDP's Water Program is produced for 470 investors representing $50 trillion in assets, and is based on information submitted to CDP by 185 Global 500 companies.[20] The results from the CDP for Water 2012 report, prepared by Deloitte, reveal that 53% of respondents have experienced negative impacts from water-related challenges, including water scarcity, rising compliance costs, regulatory uncertainty, and poor water quality over the past 5 years.[21] Institutional investors have the ability to use both the Investor CDP and CDP for Water programs as tools to evaluate a business's climate and water risks as a factor in their investment decisions.

The aim throughout CDP's programs is to use market mechanisms to influence the spread of environmental and social transparency into organizational processes, first through large corporations and cities, driving the practices into all business metrics. The process for creating a CDP response is as simple as sending the CDP an email requesting participation.

Dow Jones Sustainability Index
and Newsweek's Green 500

There are other indices and frameworks that firms pursuing more sustainable practices should be aware of. The Dow Jones Sustainability Index (DJSI), an invitation-only index using a Corporate Sustainability Assessment questionnaire that runs over 70 pages, is targeted at the largest 2,500 companies listed on the Dow Jones Global Total Stock Market Index. The DJSI is also industry specific in that it evaluates companies based on 58 different industry classifications.[22]

Evaluations are weighted by using three evenly scored categories, called dimensions, which include economic, environmental, and social criteria. To make the list, companies must demonstrate improvements in their long-term sustainability plans as well as the corporate sustainability assessment questionnaire, company documentation, media and stakeholder reports, and personal contact with the companies.[23] Scoring is compiled on a variety of factors under each dimension based on the information provided by each company.

Other rankings are based on a performance evaluation in which scores are aggregated from multiple sources. Each year the magazine Newsweek, in partnership with Trucost and Sustainalytics, issues their Newsweek Green 500.[24] A "Green Score" for each firm is derived from three components:

- Environmental Impact Score (45% of the total) compiled by Trucost, involves over 700 metrics—a comprehensive, quantitative, and standardized measurement of the overall environmental impact of a company's global operations.
- Environmental Management Score (45%) compiled by Sustainalytics, is an assessment of how a company manages its environmental impacts, including the environmental footprint, policies, programs, targets, and initiatives of both its own operations and its suppliers and contractors. These are measured through a GHG Inventory. Also included are the impact of its products and services which are measured by LCAs.
- Environmental Disclosure Score (10%) evaluates the quality of company sustainability reporting and involvement in key

transparency initiatives such as the Carbon Disclosure Project and the next framework to be examined, the GRI.

The CDP programs, GHG accounting, and LCAs are all commonly used tools to evaluate environmental risks and consequences. Remember though, that sustainability goes beyond addressing economic and environmental concerns to include the social component. The GRI is one framework that more intensively integrates the social aspect and can be useful for any organization.

Global Reporting Initiative

Pressure is mounting on firms to produce an annual sustainability report. Media, institutional investors, and activist shareholders aside, when a company's competition and their customers are issuing reports, they are seen as well behind the curve.

The DJSI and the Newsweek Green 500 are among the more well-known avenues through which companies are scored on their sustainability report, although there are now many lists and awards for sustainability across the commercial spectrum. Evaluation reflects whether a sustainability report follows a methodology, the most popular of which is arguably the GRI. GRI was created in 1997 by Ceres, a network of companies, investors, and public interest groups, and the Tellus Institute, a non-profit research and policy organization, along with the support of the UNEP. GRI's mission is to "make sustainability reporting standard practice by providing guidance and support to organizations."[25]

A non-profit organization, GRI provides organizations with widely accepted and utilized standards for sustainability reporting. More than 4,000 organizations from 60 countries use the framework to produce their sustainability reports.[26] Firms currently use the third version of the framework, known as G3.1, although a fourth version is being released as this book goes to publication. The GRI G3.1 framework will be grandfathered in until 2015. Its performance indicators aim at economic, environmental, social, and governance performance, and are divided into five categories: environmental, human rights, society, products, and economics. Table 6.1 shows the structure of the GRI framework. Each category has subcategories.

Table 6.1. Overview of GRI Reporting Issues for Corporate Responsibility

Area	Category	Aspect
Economic	Economic Impacts	Economic performance Market presence Indirect economic impacts
Environmental	Environmental	Materials Energy Water Biodiversity Emissions, effluents, waste Suppliers Products and services Compliance Transport Overall
Social	Labor Practices, Decent Work	Employment Labor/management relations Occupational health and safety Training and education Diversity and equal opportunity
Social	Human Rights	Investment and procurement practices Non-discrimination Freedom of association Collective bargaining Child labor Forced and compulsory labor Disciplinary practices Security practices Indigenous rights
Social	Society	Local community Corruption Public policy Anti-competitive behavior Compliance
Social	Product Responsibility	Customer health and safety Products and service labeling Marketing communications Customer privacy Compliance

Each subcategory has one or more core indicator reporting requirements. For example, one indicator under emissions is the total direct and indirect GHG emissions by weight. Third-party verification of reports is not required, and one criticism of the standard is that some organizations manipulate it to appear more transparent while not improving their performance. Yet, as a tool for organizations sincerely seeking to capture the rewards of integrating sustainability, the GRI G3 uses a comprehensive set of metrics that also show how commerce has systemic ramifications.

The indicators can be useful for almost any size enterprise in identifying areas to address and metrics to apply to those areas. Selecting even just one or two pertinent indicators from each category can be a straightforward way of stepping into sustainability issues. Some GRI indicator data are relatively easy to collect, such as

- direct energy consumption by primary energy source;
- total water withdrawal by source;
- total number of incidents of discrimination and corrective actions taken;
- total workforce by employee type, employment contract, and region, broken down by gender; and
- percentage of employees covered by collective bargaining contracts.

While selecting a few indicators to start with is straightforward and fairly easy, the full GRI reporting format is most often utilized by large corporations due to the resources required to gather and calculate a number of the indicators. The work of obtaining and compiling the data needed to have an extensive reputable, credible, and effective sustainability report that complies with GRI's requirements may be equivalent to one or more job positions. A number of firms hire sustainability consultants to guide them with these efforts. Several of the metrics discussed earlier in this chapter fold into a GRI report. Within the environmental category, a GHG inventory allows an organization to report on several critical indicators. An LCA is critical in responding to indicators under the product responsibility category. The incorporation of social data, tied with the GRI's request for financial data makes it the premier source for evaluating environmental social governance, TBL, and corporate social responsibility performance.

However, many private and non-profit organizations are also reporting to the GRI. In 2010, a team of students and faculty from the University of Massachusetts at Dartmouth assembled the first ever GRI sustainability report for a university. The Fullbright Academy for Science & Technology, a small non-profit organization, and San Francisco Public Utilities have compiled their GRI sustainability reports. Smaller organizations have the advantage of finding GRI-requested information more quickly while having a lower barrier of internal politics to overcome when it comes to making sustainability disclosures.

Shareholders of public corporations are using this information to drive investment decisions, as it provides them with a perspective on the impacts, risks, and long-term sustainability profile of an organization. As shareholders and insurers become more aware of these risks, their concerns force business leaders to attend to them. As such, the GRI is among the mechanisms that help organizations make sustainability reporting as routine as, and comparable to, financial reporting (see Figure 6.1).

Third-party verification of reports is not required; however it does play a role. Reports with third-party verification are given a plus sign next to their indicator level to signify this added step. Often, the GRI indicators serve as a starting point for organizations that just want to report, but don't fully understand what it takes to actually complete a report.

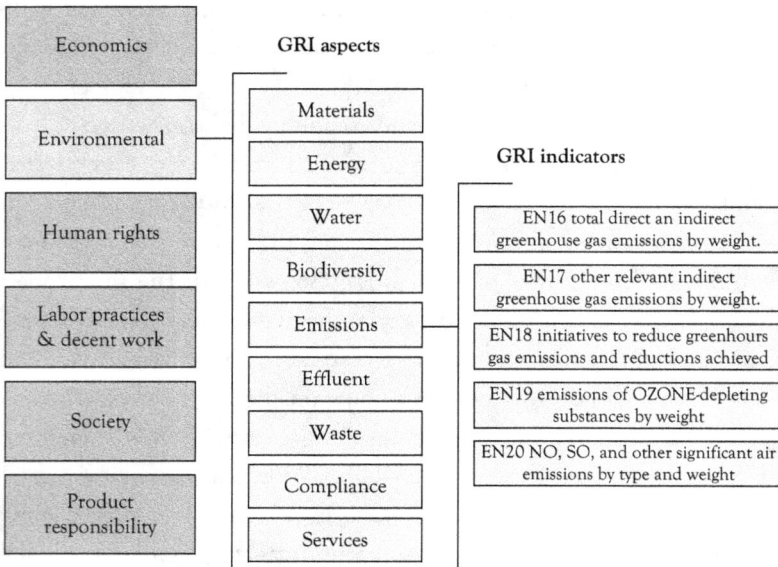

Figure 6.1. GRI G3 categories, with an example of aspects and associated indicators.

The GRI's incorporation of economic and social data is what makes this framework powerful enough that the idea of integrated financial and sustainability reporting has grown into a pilot phase. In early 2013, GRI and the International Integrated Reporting Council signed a new memorandum of understanding committing to the idea of integrated reporting.[27] Along with the International Integrated Reporting Council, the Sustainability Accounting Standards Board is focused on giving investors the information they need on a company's ESG performance by developing a sector-specific approach.[28] Both organizations are making progress in 2013 with pilots and the development of sector-specific working groups.

Through the integration of financial and sustainability reporting, the links between them will become more apparent and transparent to shareholders. Shareholders in many corporations have begun to put forward resolutions for the disclosure of sustainability performance through the conduit of sustainability reporting as a way to drive investment decisions. Shareholder requests for environmental and social resolutions accounted for more than 40% of all shareholder resolutions submitted in 2012, up from 30% in 2011.[29]

The example below is the result of efforts by Comcast shareholders who requested that the organization require a sustainability report in 2007.[30]

PROPOSAL 5: TO REQUIRE A SUSTAINABILITY REPORT

The following proposal and supporting statement were submitted by the General Board of Pension and Health Benefits of the United Methodist Church, 1201 Davis Street, Evanston, IL 60201-4118, which has advised us that it holds 658,209 shares of our common stock.

WHEREAS:

Investors increasingly seek disclosure of companies' environmental and social practices in the belief that they impact shareholder value. Many investors believe companies that are good employers, environmental stewards, and corporate citizens are more likely to generate better financial returns, be more stable during turbulent economic and political conditions, and enjoy long-term business success.

Sustainability refers to endeavors that meet present needs without impairing the ability of future generations to meet their own needs. According to Dow Jones, "Corporate Sustainability is a business approach that creates long-term shareholder value by embracing opportunities and managing risks deriving from economic, environmental, and social developments. Corporate sustainability leaders achieve long-term shareholder value by gearing their strategies and management to harness the market's potential for sustainability products and services while at the same time successfully reducing and avoiding sustainability costs and risks." (http://www.sustainability-index.com/htmle/sustainability/corpsustainability.html)

We believe that improved reporting on environmental, social, and governance issues will strengthen our company and the people it serves. Furthermore, we believe this information is necessary for making well-informed investment decisions as it speaks to the vision and stewardship of management and can have significant impacts on our company's reputation and on shareholder value.

Globally, over 2,000 companies produce reports on sustainability issues (www.corporateregister.com). Several telecommunications companies have already produced sustainability or corporate responsibility reports, including AT&T and Verizon.

The GE 2006 Citizenship Report provides a compelling rationale for sustainability reporting: "Investors are increasingly interested in evaluating companies based on a broader set of criteria than just financial performance. ... The strength of reputation, trust in brand and governance, and the ability to perform as a good corporate citizen, all impact GE's valuation and shape the perception of the Company's worth. In fact, according to a recent study, 70% of institutional asset managers believe the Company's citizenship factors will be part of mainstream analysis in the next 3 to 10 years. ...GE's focus is on providing transparent communications relating to the Company's citizenship performance."

RESOLVED: Shareholders request that the Board of Directors issue a sustainability report to shareholders, at reasonable cost, and omitting proprietary information, by December 31, 2007.

Sustainability reporting is, by definition, a way in which organizations assess their own accomplishments and shortcomings. In the example above, Comcast shareholders voted to reject the shareholder request for a a sustainability report in 2007. In the response, Comcast offers *"our current policies and practices concerning social, environmental and economic issues already address the concerns behind this proposal, and our current disclosure already provides shareholders with meaningful information regarding several of our activities in these areas."*[31]

Market-Oriented Indices and Certifications

Some indices and certifications are more market-based, with the customer or consumer as the driver of the motivation for firm participation. The Walmart Sustainability Index, which the firm developed in association with The Sustainability Consortium, was created to provide a way to measure and report product sustainability throughout the supply chain. Walmart's supplier assessment, discussed in the previous chapter, drives sustainable practices deeper into the industrial processes of thousands of products in over 200 categories using a lifecycle format. The assessment includes questions relating to energy and climate material efficiency, natural resources, and people and community. The firm reports that they have increased their sustainability index score by an average of 20% in general merchandise, 12% in grocery, and 6% in consumables and health and wellness, and plans to increase that category coverage to 300 by the end of 2013, with over 5,000 suppliers participating.

Industry-specific certifications are growing, as well. Just as the building industry has LEED as one well-accepted framework for green building, the Higg Index, developed by the Sustainable Apparel Coalition, was designed as an indicator to measure the social and environmental performance of apparel and footwear products. This index is intended to reduce confusion and redundancy by being the leading apparel industry evaluation and communications venue for sustainable practices and products, and is based on Nike's Apparel Environmental Design Tool and the Eco Index.[32]

Third-party consumer products and services certification organizations are another avenue for businesses who wish to jump on the certification bandwagon to strengthen customer trust, increase brand recognition, while also integrating sustainability into their core operations and improve their efficiency and quality. Green Seal, started in 1989, and EcoLogo, established in 1988, both develop lifecycle-based standards and work with independent verifiers of those standards.

Sustainability in Business: Moving Forward

Despite the failure of the shareholder proposal discussed above, it reinforces that global organizations struggle with what information, both positive and negative, to disclose publically. Boards and C-suites at some of the world's largest companies continue to debate whether or not sustainability should be a top strategic priority, and if so, what and how much to report. Disclosure creates transparency and generates expectations; not all companies are eager to raise the flag of sustainability just because it appears to be the next big driver in business. Organizations need the case for sustainability to be spelled out with hard data specifying the return on investment. While sustainability appears to be similar to any other organizational function, a competing interest that requires resources, time, and data, this and previous chapters have made the business case for its necessity.

The *State of Green Business 2013*, released by Greenbiz and Trucost, reports that "most companies now disclose at least some environmental impacts, and a growing number are having third-party assurance completed on their quantified performance data to make their reporting more credible."[33] The study emphasizes the need to decouple economic growth from environmental damage in the "new normal" state of resource constraint and economic volatility.

The report also notes that U.S. firms are behind the curve, weakening their global competitiveness. The consistent key performance indicators used by firms for the environmental component of sustainability, included 41% reporting GHG Emissions, 27% focused on Water (disclosed with GRI and CDP for Water and measured as a Scope 3 Emission), 7% on Acid Rain (captured in an LCA), 5% for Dust and Particles, and 20% Others. The first four categories represent 80% of the overall measured environmental footprint.[34] Collectively, these key performance indicators and the tools presented in this chapter represent some of the knowledge organizations need in order to embrace sustainability systemically.

The sustainability landscape is still shifting and developing, and reporting formats are continually being refined, improved, and deepened, such as with sector-specific reporting frameworks. They provide a standardized, level playing field for benchmarking, measuring, monitoring, and reporting progress. Growth in the use of these frameworks and reporting tools has been strong, despite the global economic downturn.

This chapter has briefly introduced an assortment of the more popular tools currently empowering organizations with meaningful metrics, tailored to their business's most material environmental, social, and economic impacts. Having such reporting formats and indices can be the difference between an effective and credible sustainability plan and a greenwashed band-aid.

Chapter Summary: Key Takeaways

The pursuit of sustainability is not a one-size-fits-all proposition; it must be tailored to each organization according to its size, industry, values, and capacities. Some of the more common resources for measuring and implementing environmental impact and change strategies include inventorying GHGs, conducting lifecycle analyses, or utilizing a standard reporting framework such as that offered through the International Organization for Standardization and CDP's programs. The GRI offers a framework that integrates not only environmental issues, but also social, economic, and governance components.

While full use of these tools, standards, and frameworks are the province of larger organizations, they offer smaller and mid-sized firms plenty of guidance to get started with metrics. The use of any of them also refocuses measurement to align with a more systemic understanding of the interface of industry with the larger systems of society and environment in which it operates.

CHAPTER 7

Where to Start in Your Business

The previous six chapters provided a concise outline of an upgraded mental model, one that includes yet surpasses previous worldviews: a systems approach that recognizes the interdependent nature of commerce with society and the biosphere. Concern for sustainability, a systemic concept ensuring the ability of both current and future generations to meet their needs, has become obvious and serious enough that it has moved beyond the province of tree huggers and radicals. An overview of the background regarding the science behind systems and sustainability was provided, as well as its significance for people on the personal and individual level, organizationally, and on the broader scale of industry.

Sustainability matters; hopefully, by this point your mental framework as well as your enthusiasm for integrating it into your organization has been shifted into high gear. This final chapter concludes with ideas to get readers started with the "how." This chapter is intended to be a preliminary overview of information to provoke thought, conversation, and planning for further inquiry into sustainability practices in business processes and models.

What You Need to Know to Get Started

Resources You'll Need:

- A clear grasp of and facility with the business case for sustainability as it applies specifically to your firm
- An ability to connect the integration of sustainability practices with your firm's strategic objectives

- A small group of sustainability champions within the organization to share the task of planning, networking, coordinating, and nudging the effort forward
- The skill of engaging with those who may initially disagree with you to discover their concerns and needs so that you can address them and tie them into the sustainability plans
- C-suite and board support or, minimally, support from at least one or two influential C-level executives or board members
- A plan to engage stakeholders in your sustainability plans, or at least to get an adequate level of buy-in
- Sufficient funding to engage in low-risk projects to prove the value of further sustainability projects
- A framework and a set of tools (or a consultant who will provide these) to guide your plans
- Persistence
- Emotional resilience
- Mental flexibility
- Patience

Overview of Sustainability Planning Action Steps

In earlier chapters, we established why industry has the potential to be the biggest contributor to sustainability solutions. Those firms approaching the opportunity from a reactive mode in which they are simply addressing regulatory, media, and public pressure, and even those making more substantive incremental changes are likely to see only a marginal payoff for those efforts. In contrast, firms with a more radical approach have reached new markets, developed the products and services in the next generation of enterprise, and are reaping the rewards, while having a larger positive impact. That said, in any organization it will take some time, effort, patience, and support to transition from concept to practice. In their book, *The Step by Step Guide to Sustainability Planning*, Darcy Hitchcock and Marsha Willard outline a practical set of steps:[1]

- First, establish the company-specific business case for integrating sustainability into the organization and, using the company's mission, develop a vision based on this business case.

- Second, choose or design a framework (discussed in Chapter 6) that fits for the industry, company, and vision.
- Third, conduct an assessment of the organization's sustainability impacts to baseline them and prioritize the most promising programs.
- Fourth, using the chosen framework, impacts assessment, and long-range plan, devise a set of metrics to track progress, and a set of sustainability strategies.
- Fifth, implement the chosen programs, while also training employees, setting up management systems and structures to ensure actions, as well as ongoing monitoring and reporting.

Diving in

All sustainability initiatives, like many organizational functions, go through a cycle of planning, implementation, monitoring, and review, where adjustment and improvement is ongoing not only within each cycle but also from cycle to cycle. The programs that come out of the planning process may vary in monitoring time scale, and every program review provides an opportunity to tweak plans, implementation, and monitoring for the next cycle.

A bottom-up approach to sustainability efforts can certainly work, but will have the most success when those in top leadership roles champion them. Support at that level will ensure that such efforts are given the

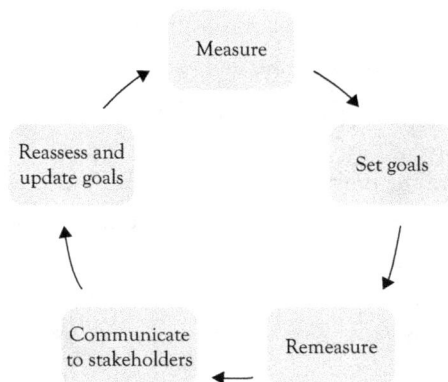

Figure 7.1. The sustainability process cycle.

human and financial resources and the prioritization that will allow more speed and ease in research, implementation, and continual monitoring and reporting. Building a strong business case for these efforts, as laid out throughout this book, helps create buy-in on the part of boards and executives.

Tie the business case for a sustainability initiative into the firm's mission and create a 10- or 20-year plan with ambitious goals. Some examples of ambitious goals are

- company-wide zero waste;
- carbon neutrality;
- 100% renewable energy sources;
- closed-loop production;
- workforce gender parity;
- complete value chain compliance with geographically adjusted living wages and safe working conditions; and
- humane working conditions.

Backcast from those goals, creating 15-, 10-, 5-, 3-, 2-, and 1-year goals that, if accomplished, will result in the achievement of the long-range goals. This process will determine what will be measured, how it will be measured, how often it will be measured, and how often it will be reported.

As Bob Willard explains in both *The Next Sustainability Wave*[2] and *The Sustainability Champion's Guidebook,*[3] be aware that appealing primarily to moral considerations without addressing payback to the organization can generate more resistance than endorsement.

Even with top-level support, it will take time and patience to

- understand the most salient risks for your business;
- uncover and research approaches to capitalizing on the most promising opportunities;
- gather information from internal and external sources, including stakeholders;
- communicate plans;
- baseline current practices;

- establish metrics and procedures for ongoing monitoring and reporting;
- train and assist employees and value chain partners as they implement programs.

Programs that are led by a purely volunteer force can be successful in the short run but risk losing momentum through attrition, frustration, and lack of time to dedicate sufficient attention. For that reason, committing funding by creating a position to manage the workload or formally assigning the tasks as part of an existing position will offer better traction. One of the roles of such a position is to develop and implement a plan to quantify what may seem to some within the organization as intangible benefits.

Be prepared for pushback, and be willing to listen to and address the concerns of those who are not yet convinced that sustainability warrants organizational priority. Executives may advocate programs, but lower lever management may create roadblocks because of a perceived dictate, conflict of priorities, or lack of urgency. Other stated strategic objectives may appear to be in competition, but just about all of them can be made easier through sustainability planning. It is likely that sustainability initiatives of some kind are already embedded within existing business practices. Existing key performance indicators can serve to identify opportunities into which sustainability initiatives can tie, in ways that may contribute to time savings versus additional work.

If support is low, start with projects that have a faster measurable impact and a quicker return on investment. Internal practices and processes that reduce overhead, such as efficiency and effectiveness improvements, discussed in more detail below, are a good place to identify suitable projects. Unless stakeholders demand a sustainability report, to manage expectations communicate programs and metrics only internally until they are established and measurable results can be documented.

How to Think Systemically About Change

Management experts have plumbed systems science to explore the most productive ways to effect sustainable change in organizations, resulting in a set of leverage points that provide a streamlined set of guidelines for strategizing change.[4] The most potent leverage points, from greater to lesser strength, are as follows:

- The ability to understand the paradigm through which a problem presents itself, and the capacity to be flexible in adjusting the existing lens to one that better explains the problem sources versus only the symptoms. That has been the primary focus of this book!

- Change the goals (purpose) of the system. As covered in the first three chapters of this book, when the purpose of a subsystem is in conflict with the purpose of the system within which it operates, the larger system endures problems. Increasing shareholder value is, arguably, the primary goal for any for-profit business organization. Due to a human bias toward short-term thinking, that goal has resulted in a take—make—waste industrial model, spawning external costs that damage overarching social and environmental systems. Sustainable enterprise and the benefit corporate structure (discussed in Chapter 5), and moving from a product to a service model are examples of changing the system goals so that they are better aligned with social and environmental system purposes.

- Change the rules of the system. The mounting evidence of the economic, social, and environmental consequences of human resource consumption, which is aided by industrial processes and practices, has begun to usher in a shift in industrial rules. Institutional investors, insurers, governments, and customers have become increasingly aware that a share of those consequences is the result of firms creating an artificial boundary of responsibility and a focus on short term profit making. This realization is causing a shift in rules about what is included and excluded in a firm's consideration, and delegitimizing the accomplishment of profits when they incur costs to society. Rules include policies and procedures as well as the unwritten cultural rules of the organization ("That's not the way we do things around here!").

- Improve access to information—not just any information, but the kind that promotes knowledge that increases both understanding of the business as a system and its wise application. Expanding access to information beyond departmental or team silos results in not only greater accountability, but improves the speed with which problems can be identified, analyzed, and resolved. Doing so also reveals the need for more or better information to assess current performance and design intervention strategies to improve efficiency and effectiveness, reduce negative impacts, innovate new services or products, or develop new markets. Sustainability reporting is an example of this leverage point.

All-Too-Human Pitfalls to Avoid

As with any effort aimed at improvement, plans must address resistance to change. Aside from the tendency to think in the short term instead of planning and considering strategies that serve a more extended span of time, there are other human inclinations that act as barriers or detours to change efforts.

People prefer to maintain their situation rather than change it for something unknown even if the change is likely to result in improvement, a decision bias known as loss aversion or, more colloquially, as "the devil you know is better than the devil you don't."

There is also the human inclination to take mental shortcuts using existing knowledge and beliefs, even if they aren't relevant to the situation, or gut responses that are at odds with empirical evidence. This results in ineffective decision making and metrics that may be too simplified to result in the kind or degree of change sought. Another cognitive error is conflating wants into needs; the latter are non-negotiable, but the former are simply desirable.[5]

In spite of these inclinations, there are ways to sidestep them:

- From the start, devise decision-making criteria that incorporate a longer time frame.
- Incorporate feedback and engagement from a representative set of internal and external stakeholders.

- Set ambitious long-term goals and then backcast them into interim goals.
- Create monitoring and reporting criteria, expectations, and frequency, as well as who will be responsible for gathering these metrics and creating and distributing the reports.
- Communicate the goals and progress widely and often, internally as well as to all stakeholders.
- Make the default (easy, obvious) option the sustainable choice in all policies.

Beyond these psychological and sociological components, which are needed to think more systemically and knowledgeably about intentionally integrating sustainability, are the quantitative nuts and bolts of the frameworks and metrics discussed in the previous chapter. They are explored here through the lens of efficiency and effectiveness.

Efficiency

Efficiency, especially radically increased efficiency, is key to sustainability efforts since many programs will involve the use of resources: material, energy, human, and financial. Industry best practices are one means to establish a comparative standard. Line employees and management may have viable ideas for improvements. Below is basic information aimed at increasing efficiency using some of the standards and metrics discussed in Chapter 6. First discussed are the avenues that address environmental management systems, with less focus on the explicitly social side. The more balanced frameworks which incorporate broader and more focused social aspects are included in the later sections.

The Basics of Measuring Carbon

Carbon is correlated with efficiency in the use of energy, one of the largest impacts for most businesses. Every oil-derived vehicle mile, whether by air, train, truck, or car, freight or passenger vehicle, every kilowatt hour and cubic foot of natural gas is part of an organization's carbon footprint. Greenhouse gases, including carbon dioxide and the other primary GHGs, such as methane, nitrous oxide, hydrofluorocarbons, perfluorocarbons, and sulphur hexafluoride, can be thought of as a set of efficiency metrics.

In a typical facility, GHG emissions are measured in categories of decreasing direct influence or scope. Scope 1, or direct, emissions are those that are emitted from a pipe or tower that the organization owns. Scope 1 emissions originate from energy sources such as natural gas for heating or creating process steam, diesel or unleaded fuels used to power company-owned trucks or cars, liquid propane for forklifts, and the refrigerants used to run air conditioning units. Certain industry sectors produce additional GHG forms. For example, dairy, beef, and other livestock businesses produce methane. The breakdown of fertilizers on the landscaping of golf courses and on agricultural lands is also considered Scope 1. The Scope 1 category comprises emissions in which there is visible emission, such as from a tailpipe, smokestack, or burning in a boiler.

Scope 2 emissions are often referred to as indirect emissions, implying that the actual emission is happening at a place one level removed from the firm. Electricity is among the most prevalent source of Scope 2 emissions. An electrical or electronic device, when plugged into an outlet, does not result in an emission at that location. The combustion of coal, nuclear reactors, hydropower, or other power production methods occurs at the power production facility. Scope 2 emissions also include heat, steam, or chilled water purchases from an external provider. Emissions from charging electric vehicles are another example of Scope 2 emissions.

Unlike Scope 1 and 2 emissions, Scope 3, known as supply chain emissions, are generally not requested by most reporting schemes like the GRI and CDP. However, Scope 3 emissions still represent emissions for which an organization is responsible, even though they do not have direct control over these emissions.

Employee commutes are included within Scope 3 emissions. As the company requires employee onsite attendance, it is exerting influence over that carbon-intense behavior and is thereby indirectly involved in the generation of those emissions. Incidentally, commuting is also a personal Scope 1 emission of the individual employee, whether the commute is by public transport, a pick-up truck, or a biodiesel vehicle.

Delivery services such as FedEx and UPS fall under Scope 3 emissions. The organization utilizing such services to send packages is responsible for the emission since they create the need for the service, under Scope 3. Both FedEx and UPS offer programs for large users to collect Scope 3 emission data, based on the number of packages sent using a specific account number. UPS has detailed data, tracking every package by plane and vehicle type, total mileage, and annual fuel economy of the actual vehicles employed in the delivery process.

Counting emissions this way may appear to result in double and triple counting. Because the generation of an emission usually involves two or more parties, the three-scope system within GHG accounting allows for these emissions to be allocated to different groups based on scope. This framework also drives collaboration, allowing suppliers and customers to discuss emission reduction strategies including improved delivery methods that drive down emissions, thereby improving efficiency of a product or service.

Greenhouse gas emissions can correlate with the inefficient use of operating expenditures. Lower emissions mean a more efficient use of resources. For example, lighting and heating, lighting, and air conditioning use in unoccupied spaces is a form of wasted resources and unnecessary emissions. Leaving computers, printers, and other devices and equipment on when not in use generates emissions while also wasting kilowatt hours. A pound of waste is a source of emissions at the landfill but also has costs associated with disposal, and is also related to a raw material purchase. Idling a vehicle while not logging miles results in wasted emissions, as well as wasted fuel and a higher operating cost per mile.

Matching emission data to production or operational data serves as a business management tool. Some firms collate data to determine total emissions per product, total emissions per customer, emissions per hour of operation, or emissions per employee. These are useful key performance indicators for monitoring performance normalized to output rather than a simple absolute number. Greenhouse gases are a potent baseline to start a conversation about sustainability with colleagues who may not understand or embrace sustainability as an organizational imperative.

The Basics of Measuring Lifecycle Assessment

Introduced in the previous chapter, LCA is another measure of efficiency across a product or service. The use of this tool as a starting point provides the foundation for how progress in sustainability translates throughout an organization's value chain, defining efficiency as part of that value chain.

Lifecycle assessments comprise three key stages. The first is "cradle to gate," which includes all of the raw materials, products, and processes that occur before those items reach a business's facilities. As an example, the assessment of a traditional button-down dress shirt might include its polyester and cotton fibers, sewing thread, dyes or dying process of the fibers, all of the processing of the natural and synthetic fibers, buttons, and tags. A cradle-to-gate LCA for pet food includes grains, proteins, digestion aids, and other supplements, including water, fertilizer, pesticides, tractor and transportation fuels, and other sourcing impacts for each ingredient. Each element of the value chain comes from the range of processes representing varying levels of impact, data that isn't visible without a tool like LCA.

The second stage of LCA is "gate to gate," representing everything that happens within a firm's physical plant or operations. In the case of the dress shirt, it includes all processes involved in the sewing and finishing of the garment, and its packaging. The gate-to-gate phase represents the electricity usage associated with sewing machines, the trimmings and waste associated with cuts, stitching, packaging, and other processing. Depending on the particular assessment model used, calculations may include economic factors such as time, labor, safety issues, and other fixed or variable costs.

The final phase of LCA is "gate to grave," which includes all aspects of the product or service once it leaves a manufacturer's operation. The dress shirt moves in stages: from a plant to a distribution center, from there to a storage facility, onward to a store, purchased and brought into a home, where its use and disposal are also considered. From the first wash to the last, including the water, detergent, phosphates, and electricity to wash, dry, iron, and fold, with varying factors that change if the item is light or dark in color, and beyond, to the landfill, LCA computes all of these impacts.

The LCA has been developed into a standard science and can be applied to any industry from electronics, to dairy, and beer manufacturing, to services such as insurance, banking, or telecommunications. Every step in the value chain comes with its own measurement of efficiency. Process improvement strategies driven by LCA data identify opportunities for adjustments that result in lower impacts while increasing efficiency in product design and development.

Lifecycle assessment data can be utilized to compute metrics such as waste per pound of product produced by process and stage, embodied energy of raw materials per kilogram, water consumption by finished good from cradle to gate or total GHG emissions per pound of product produced or per customer subscribed, visited, or engaged. By establishing a baseline figure, it will be easy to determine whether improvements in efficiency are being realized.

Low-Hanging Fruit

There is a reason that a discussion of the programs that produce easy and quick returns appears so far into the topic of efficiency. Too often, these types of programs are scattershot and are not accompanied by an organizational review and development of strategies that will offer substantial reduction in impact. There is a difference between the business-as-usual approach to being only a bit less unsustainable and taking actions that factually and substantially bring impacts down to a level closer to the planet's constraints. The point of corporate sustainability is not simply for the corporation to sustain itself, but to effectively reduce industrial impacts so that the planet's and societal systems can still provide the services vital to human existence.[6] See our discussion on effectiveness, later in this chapter.

The low-hanging fruit does provide an opportunity to get some low-risk experience, generate some momentum, and build stronger buy-in for other programs. In the arena of sustainability, conversations around such opportunities to improve efficiency are often followed by a discussion of lighting, which starts with an inquiry into current lighting equipment condition and the speed with which a return on investment can be accomplished. Other examples of low-hanging fruit include recycling, such as the monetization of a waste raw material like cardboard, aluminum, or some type of plastic. Office paper and office waste fall into this category as such efforts help contribute to a reduction of material sent to a landfill. Most of these sorts of projects are not accompanied by analysis of data because they simply seem to make sense. Lighting technology improves, waste can be monetized, and office paper should always be recycled if not avoided.

That's not to say that a firm should ignore these opportunities, only that they will serve best if they are part of a wider organizational assessment that determines the firm's most significant negative externalities. Once done, the easy and obvious programs are completed in the context of a plan to drive sustainability much deeper into the organization, even if that deeper integration takes years.

Easy programs can yield surprising results, if they are done well, but return on investment may be harder to come by if adequate research is skipped. For instance, one company replaced their bulbs but not their ballasts, leading to a lower-than-expected efficiency. Identifying simple errors in systems makes an excellent starting point for capturing quick wins. This is an area in which a sustainability consultant can be a worthwhile investment. A consumer product goods company brought in a sustainability consultant who identified overpaid utility taxes occurring during a period of 3 years, netting the company over $650,000 in rebates from the state tax commission. Another firm had been paying utility bills for over three-dozen facilities, ranging from 10,000 to 150,000 square feet, that they no longer owned; their savings exceeded $1 million. A firm with an excellent waste recovery and recycling program was being paid below market values for their highly valuable waste stream. Over the last 5 years they have reaped an additional $2 million per year in addition to the $500,000 they had been receiving, an excellent return on their investment in a contract with a sustainability consultant. A municipal project identified over 800 streetlights for which municipal taxes were being paid that were no longer in operation. At the cost of $28.08 per month per light pole, this resulted in overspending of nearly $270,000 per year for the municipality.

Programs that qualify as low-hanging fruit require more than just a good idea; they can become new platforms for broadening and accelerating sustainability enterprise wide, but it all starts with data.

Experimenting with the GRI and CDP

Efficiency and sustainability also have meaning on the social indicators. As tools, the Global Reporting Initiative and CDP offer opportunities to capture data to improve efficiency on the human side. A 2009 study found that employees in green buildings are more productive than those who work in non-green buildings. The study of 154 green buildings and 2,000 tenants found that 45% of respondents reported that they had experienced an average of 2.88 fewer sick days at their new, green office location versus their previous non-green office location.[7]

Healthier employees are more productive and are certainly more engaged. Social performance indicators included in the GRI's G3 framework relating to labor highlight happiness factors such as employee turnover and benefits for full time employees. Organizations with high turnover almost always experience efficiency problems. The cost of low employee morale can be significant. The Gallup Organization estimates that there are 22 million disengaged employees costing employers $300 billion annually.[8] Research published in 2010 by Warwick Business School indicated that happy workers performed 12% more productively than the control group.[9]

Attention to some GRI indicators can assist a firm in avoiding future discrimination lawsuits: equal remuneration for women and men requires reporting the ratio of basic salary and remuneration of women to men by employee category and by significant locations of operation. A study by the UCLA-RAND Center for Law and Public Policy reports that the average litigation cost for a discrimination suit at trial is $150,000, which does not include any settlement amount or the loss in productivity of management distracted by such a suit.[10] In addition, discrimination lawsuits affect a company's culture, employee engagement, and turnover, and present a risk to corporate reputation, brand equity, and sales.

Effectiveness

Carbon accounting, LCAs, and the GRI framework offer a set of possible starting points that organizational sustainability leaders need to drive operational changes. These operational changes include improving efficiency while reducing operating expenses. Both hard and soft costs can be better managed when sustainability analytics are combined with financial analytics. However, the data will require some extra attention to make sure it is effective for making decisions.

To drive effectiveness, each tool has a set of principles that need to be followed to create usable data, with well-designed data inputs generally resulting in useful data output and vice versa. The challenge in using sustainability data is that the process of creating good data is still in its adolescence, uncoordinated, and full of blemishes. To address this, tools in the sustainability toolkit come with protocols that help establish guiding principles that define quality.

Five principles of GHG management were established as part of the Greenhouse Gas Protocol Corporate Accounting and Reporting Standard, and they apply well beyond carbon accounting: relevance, completeness, consistency, transparency, and accuracy.

Relevance helps organizations define what is important to include and exclude in a GHG inventory, delimited by a boundary. The boundary can be based on the business goals of the company, financial boundaries, or operational (control) boundaries. Establishing those boundaries delineates exclusions and inclusions so that they can be explained and reported. Relevancy is a key criterion in judging the credibility and intent of an organization. For example, consider the intent of these two companies: one firm wanted to leave a significant source of emissions off their inventory because they didn't want it to taint their image; another firm, a cable company, wanted to include cable set top boxes at their customers' homes in their inventory because they "felt responsible for those emissions."

The principle of completeness defines how comprehensive and representative the compiled information is, in relation to the organization. This principle drives organizations to make informed estimates where actual data is not available rather than just ignoring the emission source (or other impact) because data was not available. It pushes organizations to make their best attempt to provide inclusive, consistent, and accurate data. In the commercial real estate sector, many leases in multiple tenant properties, particularly older ones, include utilities in the rent. This is usually the case with large office buildings or facilities that are not equipped with separate metering for each lessee. Such arrangements present a challenge when attempting to allocate the energy used by each lessee. To solve this, the Greenhouse Gas Protocol provides guidance on how to estimate the energy, water, and waste emissions within the leased spaces, based on building type and total square footage. This method is a more complete method than just ignoring leased properties because utility data are unavailable.

The consistency principle is the ability of an organization to measure emissions in a repeatable and comparable manner over time. The challenge of the consistency principle is that it forces practitioners to spell out their own methods clearly so that any other practitioner could follow those instructions and compute the same result. This includes, using consistent data sources, calculation methods, boundaries, and scopes. Application of the consistency principle can be complicated by changes to systems. Data management practices within organizations are fluid, frequently being upgraded or integrated, bringing both improvement and degradation of data. Changes to businesses also are cases in which the consistency principle is employed. As business units or divisions are purchased or sold, as operations are internalized or outsourced, or as production processes change there are also changes to emission inventories. The ability to identify, isolate, and adjust GHG data is critical to having a consistent inventory that is useful over an extended span of time.

The transparency principle provides an audit trail of the information used to create the inventory. The auditing and verification of emissions is an emerging practice. All four of the Big Four accounting firms have divisions that are focused on providing assurance and verification services on carbon and sustainability reporting data. To provide this service, organizations must be transparent about the processes and procedures used to calculate GHG emissions. All assumptions or limitations identified in the process are clearly spelled out so that they can be evaluated. The transparency principle prompts organizations to be open and clear in how they arrived at a final GHG emission calculation. A failed transparency example would be an organization that justifies not reporting its calculation methods with the reasoning that to do so would be a breach of proprietary information or intellectual property. Some organizations using GHG accounting software omit calculations or descriptions of emission factors, others fail to provide supporting evidence, like organizational charts, financial documents, or other operational information through which business units or business processes were disguised or hidden.

The accuracy principle, inherently linked to the completeness and relevancy principle, describes the precision of the organization's GHG inventory and describes how the end numbers are true and fair. If all relevant emission sources are accounted for and data for all of those sources are complete, the only remaining issue is accuracy. Accuracy is also a goal as systems have an inherent error built into them. The cost of 100% data accuracy far exceeds the cost of 90% accuracy, but the ability to realize and account for the degree of accuracy of your data actually provides a window into areas for operational enhancement.

For organizations of scale, the accuracy principle is a continuous improvement goal. Few organizations, given their dynamic business environment, are able to capture emission data at 100% accuracy. However, organizations are able to define accuracy in relation to their operations and to design processes for improving that accuracy. Similar to Generally Accepted Accounting Principles like materiality and conservatism, the spirit of GHG accounting provides a representative picture of the organization's emissions based on emitting sources that its operations were responsible for creating. That picture should be based on a realistic and conservative approach to material sources of emissions.

Global Reporting Initiative Principles

The process of creating a sustainability report using the GRI framework is also guided by a set of principles to ensure effectiveness: content, quality, and boundary.

The content principle, similar to GHG inventory relevancy, helps organizations define what should be included in and excluded from the report. The principle is fairly subjective in that different stakeholders may have differing expectations for sustainability report content. For example, if Starbucks didn't report on environmental and social issues associated with the growth and harvesting of coffee beans, then it would be viewed as missing significant content in the eyes of some stakeholders. The content principle comprises four subprinciples: materiality, stakeholder inclusiveness, context, and completeness.

Materiality guides reporters to ensure that report topics have been defined systematically and objectively. Many organizations design a materiality matrix that represents issues that are significant according to stakeholder perceptions, as well as significant in terms of quantity. Energy, for instance, is both an issue of stakeholder concern and a significant source of emissions for the organization.

Stakeholder inclusiveness requires that reporters define their stakeholders and explain how their interests and expectations have been addressed in the report. In order to understand the needs of stakeholders, they must first be engaged, either through stakeholder engagement surveys, annual stakeholder meetings, focus groups, community meetings, or one-on-one meetings with key stakeholders.

The principle of sustainability context asks organizations to define how the content fits into the wider view of sustainability. It serves to align the purpose of sustainability reporting as disclosing information about the organization's comprehensive sustainability profile.

The purpose of the subprinciple of completeness is to ensure that material topics have been defined, the stakeholders have been included, and the context of sustainability is present. This is an important principle for organizations that issue sustainability reports that are heavily focused on environmental issues over social issues, or only financial issues with little focus on environmental or social issues, to attend to.

The GRI framework also has a set of six subprinciples for establishing reporting quality. These include balance, comparability, accuracy, timeliness, clarity, and reliability, of which the first three are arguably the most useful in driving effective reporting.

Balance is a key principle in having an effective sustainability report, ensuring that sustainability is being perceived as a systemic organizational issue. It also ensures that internal and external expectations have been identified and managed through the reporting process.

Comparability serves as a management tool to gauge improvement over time and as a way to benchmark against an industry standard. Most organizations that request sustainability information from suppliers are looking for continuous improvement, over a specific target or objective. The same improvement expectation is true of sustainability reporting; a series of comparable reports over time is required.

The subprinciple of accuracy is similar to that described above for GHG inventorying. All other principles and subprinciples become irrelevant when a report is based on inaccurate data.

The accuracy principle requires data to be sufficiently accurate and detailed for stakeholders to be able to assess performance. Compliance entails being able to explain the degree of accuracy, but does not require reaching a level of 100% accuracy. Accuracy has limits and requires transparency.

The boundary principle delimits what to include and exclude based on a boundary defined by degree of control. For areas in which the organization has high control and influence, reporting on those issues should be included. Correspondingly, issues in which management has little control or little influence should be left outside the boundary.

Collectively these guiding principles help steer organizations toward effective sustainability reporting. The reporting process is an effective tool that helps drive sustainability in organizations by engaging material issues as well as the interests of stakeholders. The report serves as an a la carte menu of sustainability aspects that have been reiteratively defined by the GRI and selected through a systematic process that focused on inclusiveness and transparency.

Driving efficiency and effectiveness in an organization's sustainability program requires a systematic and systemic examination into an organization's social, environmental, and financial impacts. The tools introduced in Chapter 6 and discussed in more detail above offer avenues to consider, select, establish, and delineate the quantitative and qualitative data needed to drive effective sustainability decision making. They provide standardized and transparent methods to compare progress over time.

While there are many other tools that will help accelerate sustainability on specific tasks, the frameworks, standards, metrics, and indices introduced above represent those that best lay the foundation for managing sustainability at a systems level perspective. These resources are continually upgraded and updated, evolving with industry, science, and shifting reporting expectations.

The Future of Sustainability in Enterprise

Businesses are beginning to recognize that negative externalities generated by their operations are now becoming a cost of doing business. The world's largest reinsurance firm, Munich RE estimated natural disasters globally cost $160 billion in 2012.[11] Of that total, 67% were suffered in the United States, leading the reinsurance firm to state that the "weather related catastrophes in the U.S.A. showed that greater loss-prevention efforts are needed." The total losses amounted to more than the annual revenues of GM, Ford, GE, or Apple.[12] If that is not attention grabbing enough, the 2011 cost of natural disasters exceeded $400 billion, more than the annual revenues of every Fortune 500 firm except Exxon Mobil[13] and Walmart Stores.[14] These incidents may or may not have been related to man-made factors, such as climate change, but firms are being nudged in the direction of greater attention to such risks.

Within its Carbon Disclosure Project Investor Response, UPS highlighted its focus on climate-related risks, even providing an example of how their business was impacted as a result of natural disasters. UPS claims "Risks related to natural disasters (such as hurricanes, tornados, floods, etc.) represent the largest potential climate-related risk category to UPS." Their report continues, "The hurricane impact on New Orleans, in 2005, is an appropriate example of how this physical and financial risk arises at UPS. UPS operations in New Orleans were promptly restored after the storm, but much of the industrial base was gone. Contingency plans were put in place to bypass the affected areas where needed, minimizing any impact to the network operation as a whole." The acknowledgment of climate risks impacting business performance is intertwined in UPS's CDP Response and evident as one of four companies who scored a 99 out of 100 on the CDP's Global 500 Leadership Index.[15]

The need for sustainability and sustainable solutions continues to grow. The United Nations Population Division predicts in its "low-variant scenario" that world population will exceed 9.2 billion by 2050. With this growth come more than 2 billion new consumers of the earth's finite resources.

The physical demands this planet will endure will require a radically different level of focus and energy devoted to the discipline of sustainable design. The pressures associated with climate risks will force organizations to innovate and adapt. It is simply a matter of time. Organizations that are prepared will incur fewer shocks.

The growth of the pursuit of sustainability as a business focus is one strong indicator. As mentioned in Chapter 6, 81% (405) of corporations representing the Global 500 responded to the Carbon Disclosure Project questionnaire in 2012. In the built environment, the push for more energy efficient buildings can be best seen by the growth in LEED Certified buildings by the U.S. Green Building Council. In 2012, there were more than 12,500 new certified green buildings, a sharp increase from the 2,500 projects in 2006.[16]

The GRI has seen an uptick in sustainability reporting with a 46% absolute growth in the number of U.S. companies creating GRI reports from 2010 to 2012.[17] On the local government side, ICLEI Local Governments for Sustainability continues to grow its membership reporting that members of 12 mega-cities; 100 super-cities and urban regions; 450 large cities; and 450 small and medium-sized cities and towns in 84 countries committed themselves to sustainable development.[18]

It will take a strong and concerted set of efforts on many different projects, policies, and practices by each person, by their communities and governments, as well as NGOs and businesses, to make a dent. It is very likely that the planet and its most vulnerable populations are already seeing the beginning of the impacts of wide scale systemic change from climate, overuse of resources, and income disparity.

Many local and some state governments have already established policy, legislation, and standards. Future legislation and policies on the national scale such as carbon taxes, cap and trade, or other carbon limiting bills could impact the financial positions of companies while reducing the cost of environmental protection.

Competition amongst communities and states, industries, and individual businesses for resources and services, even for basics such as water, food, and security has the potential to become a limiting factor. Governments may find themselves integrating and dissolving layers of governance structure so as to speed up decision making related to climate adaptation. Protection of headwaters, watersheds, coastal barriers, floodplains, and water tables is already occupying the attention of worried officials.

Non-governmental and academic organizations will continue to be instrumental in conducting targeted research and development, related to the health and status of the planet, and the most promising policies and practices. They have been working with governments and businesses to drive solutions from concept to execution, serving as advisors to balance maximum disruption. Researchers at the Intergovernmental Panel on Climate Change and at other climate research centers will continue to model scenarios based on population growth, resource consumption, and global GHG emissions to determine the pace of climate change.

Within industry the focus may shift from operations to service or product and supply chain. Products that are highly carbon intensive may become cost prohibitive as legislation and availability erode their price advantage. Industry will need to adapt to changes in supply chains, possibly reverting from global to localized economies. Supply chains may shorten, and products may become inherently greener without the emergence of a major green consumer segment. Industry may see a shift in availability of raw materials or regional manufacturing capacity as climate change impacts could drive the cost of electricity, infrastructure, or water to a point where firms struggle to stay competitive.

The longer the delay in integrating a set of systemically designed changes, the more severe the cost, in terms of raw financial capital, interruption of commerce, and in human suffering. As explored in Chapter 2, a growing number of scientists believe it is highly likely that the planet is now beyond the point at which climate change can be avoided, so it is a matter of the degree (no pun intended) of severity and the ability to adapt to those changes that will determine the extent of disruption.

Even if developed nations were to reduce their emissions to zero, which is unlikely to happen in the next century in any case, the output by developing nations is very likely to cause continued emission growth. Those individuals who are nearing retirement may be lucky enough to avoid grave repercussions, but younger generations may not be as fortunate.

The companies that will thrive are very likely to be firms that have used a systems framework and an internal drive to innovate and collaborate with external partners to radically improve the quality and efficiency, and decrease the negative impacts, of their processes, products, and services. By doing so, they decrease their exposure to risk, minimize resource use and costs, enhance their brand trustworthiness, and stay ahead of legislative and regulatory developments. Stakeholder engagement and corporate citizenship are likely to be deeply integrated into business processes and practices.

Beyond Growth and Profit Primacy

At some point in the not-far future, priorities may shift. The recognition that the environment is the single largest attribute that keeps society and economies functioning may move from a third-level priority to front and center along with society and economy. The belief in never-ending growth and prioritization of profit as prime metrics may decline. Sustainability would no longer be a stand-alone concept; it may dissolve into a core function of daily life, prompted by necessity, or perhaps recognition that current models and metrics leave much to be desired.

Gross domestic product (GDP) was intended as an indicator of short-term economic output, but is instead used as a barometer of economic health and social progress. It was not intended for this broader use, and it tends to rise with household debt, crime, family breakdown, and commuting time and "increases with the depreciation of machinery and the extraction of fine minerals, while failing to reflect the long-term contributions of education, research, and entrepreneurship."[19]

Attention may swing from growth to development, from market share, margins, and short-term profits to a stronger focus on resource conservation and restoration, better access to livelihoods and living wages, healthcare, education, community well-being, civic participation, leisure time, and other metrics, such as those in the Genuine Progress Indicator (GPI),[20] an alternative to GDP. The GPI views economic activity as a means to economic welfare, meaning the level of satisfaction and utility derived from consumption and other human pursuits.

Economic welfare is incorporated into the metric as the ends to achieve, versus the economic activity that GDP measures, which is only a means, not an effective measure of ends. The GPI also factors in the costs of common resources and conditions, such as unemployment, crime, commuting, a variety of types of pollution and emissions, loss of resource base such as wetlands, forest cover, and non-renewable resources.

Another index that has been proposed as a GDP alternative is the Social Progress Index (SPI),[21] a set of 52 indicators divided into three dimensions:

- Basic human needs, such as personal safety, sanitation, shelter, adequate nutrition, basic medical care, potable water, clean air

- Foundations of well-being, including literacy and access to primary and secondary education, ecosystem stability, and health and wellness
- Opportunity, such as personal rights and freedom, equity, inclusion, and access to higher education

Developed jointly by Michael Porter, a well-known Harvard professor, working with the Social Progress Imperative, an organization whose mission is to advance global human well-being, the SPI is based on studies that have found a high correlation between a variety of social indicators and economic growth.

Porter has authored some seminal work on sustainability issues in business management with his coauthor Mark Kramer, including recent work on creating shared value, a concept created to contrast with corporate social responsibility. Shared value is about improving societal issues with a business model, using enterprise opportunity to address those issues, versus the CSR model which focuses more distinctly on the organization itself and reducing its risks. The concept implicitly recognizes the interdependency of the health of both business and society, of corporations and their communities. Shared value is social enterprise, what these authors call "inclusive business strategy."[22]

Other indices positioned as GDP alternatives include Gross National Happiness, the Human Development Index, and Ecological Footprint, all of which integrate a more systemic perspective on measurement. While these indicators may be policy level metrics, through media coverage and consumer pressure they are beginning to influence business policy, models, and strategy.

In Conclusion

A systemic and effective approach to sustainability can be full of carrots without sticks, incentives that change the rules and shift organizational goals. One of them is the incorporation of sustainability performance criteria into executive and management compensation. Inclusion on the DJSI, on Newsweek's Green 500 list, or CDP's Global 500 Leadership Index are other incentives for larger corporations.

Suppliers to retailers like Walmart and Kohl's can also be rewarded for sustainability performance. Both retailers score suppliers, but Kohl's includes sustainability within its Certified Vendor program: products from these vendors move through distribution centers unaudited, allowing products to reach the store shelves quicker than those of non-certified vendors.

Clearly, the need for integrating sustainability and a systems mindset into business models, processes, and practices presents some challenges, but each is accompanied by an advantage. Together, they prompt leaders to take the actions necessary to maintain and improve quality of life, rethink current business-as-usual paradigms, and drive innovation. These chapters have presented a convincing set of arguments about the need, urgency, and imminent requirement for these changes, as well as many value propositions and opportunities inherent in these trends.

Sustainability may mean having to address a future where much of what has supported past success is no longer available or dependable, is impacted by radically different patterns of consumption and demand, or is constrained by ever increasing competition. The level and scope of required change will touch every aspect of commerce as a whole, as well as each and every business organization.

It is likely that change will occur gradually and on a scale that no one firm or even industry can tackle individually—yet exist in an environment where no common ground of understanding or priority can be assumed. A systems perspective guides the leadership function as a driver of complex interactions and relationships toward a sustainable organization and delivers great value to shareholders and stakeholders alike.

Notes

Chapter 1

1. KPMG Advisory N.V., Global Reporting Initiative, & Center for Corporate Governance in Africa. (2013).
2. Ernst and Young (2013).
3. KPMG International in cooperation with Economist Intelligence Unit (2011a).
4. KPMG International in cooperation with Economist Intelligence Unit (2011b), p. 2.
5. KPMG International in cooperation with Economist Intelligence Unit (2011b), p. 6.
6. Starbuck et al. (2012).
7. Lubber (2012).
8. International Insurance Society (2013).
9. U.S. National Aeronautics and Space Administration discussion on the subject at http://climate.nasa.gov/effects/
10. Entrekin et al. (2011), pp. 503–511.
11. Whitea et al. (2012), pp. 645–715.
12. Gupta and Sengupta (2007), pp. 12; Gressel (1999), pp. 361–366.
13. Senge (2006).
14. Senge (2008).
15. Hawken (1993).
16. Lovins, Lovins, and Hawken (1999), p. 147.
17. Meadows (1972).
18. Checkland (1999).
19. Ackoff (1981).
20. Visser and University of Cambridge Programme for Sustainability Leadership (2009).
21. Meadows et al. (1972).
22. Meadows (2008), p. 188.
23. Zimmer (2012).
24. Avorn et al. (2001).
25. van Nood et al. (2013), pp. 407–415.
26. Meadows (2008), pp. 176–177.
27. Diamond (2006).

28. Geary (2005).
29. Elkington (2005), pp. 1–14.
30. McCulloch (2009), p. 221.

Chapter 2

1. Brundtland (1987), p. 8.
2. Brundtland (1987), p. 6.
3. Brundtland (1987), p. 38.
4. Von Grebmer et al. (2008).
5. Singer and Avery (2005).
6. Raloff (2012, November 17).
7. Dawson (2004), p. 37.
8. IPCC (2007).
9. Watkins (2008).
10. United National Secretary-General's Eminent Persons Group on the Least Developed Countries (2011).
11. OECD Forum on Tackling Inequality (2011).
12. Luhby (2011).
13. United Nations Global Compact (2013).
14. Ehrenfeld and Hoffman (2013).
15. Edwards (2010).
16. McDonough and Braungart (2013).
17. Yates (2012).
18. Savitz and Weber (2006).
19. http://www.group100.com.au/publications/kpmg_g100_Sustainability-Rep200805.pdf
20. Visser (2011), p. 7.
21. Anderson (2009).
22. Resources for firms rated as top sustainability performers: Global 100 2013 top 100 sustainable corps—8 of which are from the US at http://www.global100.org/annual-lists/2013-global-100-list.html; SAM 2012 Yearbook at http://issuu.com/sam-group.com/docs/yearbook2012?mode=window&backgroundColor=%23222222 58 firms, 8 of which are US firms. The 2012 Sustainability Leaders; A GlobeScan/SustainAbility Survey at http://www.sustainability.com/library/the-2012-sustainability-leaders#.UVnhnxmG53Y; The 2012 Sustainability Leadership Report: http://www.sustainabilityleadershipreport.com/
23. Jensen and Meckling (1976), pp. 305–360.
24. Martin (2010), pp. 58–65.
25. Freeman (1984).

26. For information regarding the state of vital resources: For water: http://
www.worldwatercouncil.org/index.php?id=25; For land degradation: http://
soils.usda.gov/use/worldsoils/papers/land-degradation-overview.html; and
http://www.globalchange.umich.edu/globalchange2/current/lectures/land_
deg/land_deg.html
27. United States Environmental Protection Agency (2013).
28. Casey (2012).
29. Cone Communications (2012).
30. Kiron et al. (2011), pp. 69–74.
31. http://www.coca-colacompany.com/our-company/stakeholder-engagement
32. Sustain Ability (2007). Stakeholder resource: A concise and dense resource
to learn more about the principles and practices for successful stakeholder engagement can be found at http://www.csr-weltweit.de/uploads/
tx_jpdownloads/SustainAbility_Practices_and_Principles_for_successful_
stakeholder_engagement.pdf
33. Kiron et al. (2011), pp. 69–74.
34. Kiron (2012).
35. Roston (2012).
36. Thomas (2012, August 15).
37. http://www.safecosmetics.org/
38. Salisbury (2011).
39. Entine (2012).
40. Kessler and Ellis (2011).
41. Neuman (2011, November 24).
42. Barringer (2012, February 2).
43. Abraham (2012).
44. Brent and Nicole (2012, July 26).
45. Mahler et al. (2009).
46. Ling et al. (2007).
47. Servaes et al. (2012).
48. Matsumura, Prakash, and Vera-Muñoz (2011).
49. AICPA, CICA, and CIMA (2011).

Chapter 3

1. Burman (2011).
2. Mintzberg, Robert, and Basu (2002), p. 67.
3. McGranahan, Balk, and Anderson (2007), pp. 17–37.
4. Jensen and Meckling (1994), pp. 4–19.
5. Sarkar (2011).
6. Battles (2000).

7. Hawken, Lovins, and Lovins (2000).
8. Mair (2005).
9. http://www.microsoft.com/environment/our-commitment/our-footprint.aspx
10. Warner et al. (2012).
11. http://www.wholefoodsmarket.com/mission-values/seafood-sustainability
12. Staff (2011, September 22).
13. Hawken (1993).
14. Kiron et al. (2013), pp. 69–73.
15. https://www.se-alliance.org/
16. www.bcorporation.net
17. http://www.who.int/mediacentre/factsheets/fs313/en/
18. World Meteorological Organization (2009).
19. According to the United Nations Environmental Program. http://www.unep.org/Documents.Multilingual/Default.Print.asp?DocumentID=52&ArticleID=60
20. http://marinebio.org/oceans/ocean-resources.asp
21. Zimmer (2010).
22. Morello and ClimateWire (2010).
23. Worldwatch Institute (2012).
24. Sweetlove (2011).
25. Wolpert (2011).
26. Stiglitz (2012).
27. Diener et al. (2010), pp. 52–61.
28. Di Tella and MacCulloch (2006), pp. 25–46.
29. Dietz, Rosa, and York (2009), pp. 114–123.
30. Klugman (2011), p. 26.
31. http://www.forbes.com/2007/12/20/microfinance-philanthropy-credit-biz-cz_ms_1220microfinance_table.html

Chapter 4

1. Hawken (1993), p. 15.
2. Hawken, Lovins, and Lovins (2000).
3. Macfarlane (2013).
4. Liebenthal, Michelitsch, and Tarazona (2005), p. 1.
5. Sulzberger (2011, April 14).
6. International Council of Mining and Metals (2012).
7. Newmont (2012a).
8. Newmont (2012b).
9. Center for Constitutional Rights (2009).

10. United Nations Human Rights (2012).
11 Jegarajah (2012, January 6).
12. Currie (2012).
13. The Project on Scientific Knowledge and Public Policy (2012).
14. Culp et al. (2008), pp. 7–16.
15. Fox (2013, April 24).
16. Rajaram (2011).
17. Saunders and McGovern (1993), pp. 107–111.
18. 3M (2012).
19. Apple (2012).
20. Apple (2013).
21. Rain Forest Action (2012).
22. The Climate Conservatory (2008).
23. New Belgium Brewing (2013).
24. Nestle (2012).
25. UPS (2012).
26. Shontell (2011).
27. http://sustainability.publix.com/what_we_are_doing/what_we_are_doing.php
28. World Economic Forum and Accenture (2012).
29. International Institute for Sustainable Development. Definition: Environmental Debt. http://www.iisd.org/*susprod*/principles.htm
30. Brown (2013).
31. Walmart (2013a).
32. Gallucci (2012).
33. Merck (2011).
34. Chouinard and Stanley (2012).
35. Patagonia (2012).
36. EPA (2011).
37. Waste Management (2012).
38. BMW Group (2009).
39. Motavalli (2013).

Chapter 5

1. Brundtland (1987).
2. Crooks (2009, June 29).
3. http://www.footprintnetwork.org/en/index.php/GFN/page/world_footprint/
4. CAN International. (n.d.).
5. Earth Hour (n.d.).

6. Revkin (2009, October 24).
7. CDP–What we do (2013a).
8. The World Bank (2012).
9. U.S. EPA (2013).
10. Makower (2005, May 10).
11. The video can be viewed at http://vimeo.com/6621817
12. Sullivan and Schiafo (2005, June 12).
13. Ricketts (2010, June 24).
14. Sustainable Growth 2012 (2013, February 10).
15. Hoffman and Bazerman (2007), p. 86.
16. Post and Altman (1994), p. 66.
17. Stead and Stead (1994), p. 25.
18. Hart (2008), p. 127.
19. Natural Capital Solutions. (2012).
20. SME's set their sights on sustainability (2011, September 1).
21. Botzas (2011, January 9).
22. Musco Family Olive Co.–Environmental Initiatives (2013).
23. KPMG GLOBAL (2013).
24. Boyd et al. (2009).
25. Oikos International (2013).

Chapter 6

1. Shaw and Mozley (1962).
2. Nattrass and Altomare (1999), p. 42.
3. Macarthur (2013).
4. McLamb (2011, September 19).
5. Oreskes (2004), p. 1686.
6. IPCC (2006).
7. Automotive Industry Action Group (2013).
8. Kanaplue and Berman (2012).
9. SAIC (2006).
10. Manzini et al. (August 2004), pp. 118–134.
11. Wimmer, Züst, and Lee (2004).
12. Jørgensen et al. (2010), pp. 376–384.
13. Winston (2010).
14. CDP (2013c).
15. CDP–What we do (2013b).
16. Federal Register (n.d.) (2009, October 8).
17. CDP (2013b).

18. Walmart Corporate Sustainability–Support Home Page (2013).
19. CDP (2013a).
20. CDP (2013d).
21. Deloitte (2013).
22. RebecoSAM (2007).
23. RebecoSAM (2007).
24. Makower (2012, October 22).
25. About GRI (2013).
26. Sustainability Disclosure Database (2013).
27. Webb (2012, November 26).
28. Sustainability Accounting Standards Board (2013).
29. Fellow (2013).
30. WikiInvest (2007).
31. WikiInvest (2007).
32. Reuben (2012).
33. Makower (2013).
34. Makower (2013).

Chapter 7

1. For detailed sustainability planning the authors recommend Hitchcock and Willard (2008), from which these steps are taken.
2. Willard (2005).
3. Willard (2009).
4. Meadows (1999).
5. Kahneman (2011).
6. Ehrenfeld and Hoffman (2013).
7. This example courtesy of Norm Miller, the academic director at the Burnham-Moores Center for Real Estate at the University of San Diego and David Pogue, the national director of sustainability for CB Richard Ellis. Miller et al. (2009), pp. 65–89.
8. Harter et al. (2010), pp. 378–389.
9. Oswald, Proto, and Sgroi (2009).
10. Blasi and Doherty (2010).
11. Munich RE. (2013).
12. Financial Times (2012).
13. Exxon Mobil (2012).
14. Walmart (2013c).
15. McIntire (2011).
16. Molina (2012).

17. Bloomberg LP and G&A Institute (2013).
18. ICLEI Global (n.d.).
19. Beachy and Zorn (2012), p. 5.
20. Talberth (2012).
21. Social Progress Index (2013).
22. Porter and Kramer (2011), pp. 62–77.

References

3M. (2012). *3M Sustainability Report 2012.* Saint Paul, MN

Abraham, R. (2012). *Good corporate citizens perform well on markets.* Retrieved December 15, 2012, from mydigitalfc.com: http://wrd.mydigitalfc.com/companies/good-corporate-citizens-perform-well-markets-816

Ackoff, R. L. (1981). *Creating the corporate future: plan or be planned for.* New York, NY: Wiley.

AICPA, CICA, & CIMA. (2011). *SMEs set their sights on sustainability: Case studies from the UK, US and Canada.* UK, US and Canada: Chartered Accountants of Canada, American Institute of CPAs, Chartered Institute of Management Accountants.

Anderson, R. (2009). *The business logic of sustainability.* Retrieved December 5, 2010, from http://www.ted.com/talks/ray_anderson_on_the_business_logic_of_sustainability.html

Apple. (2012). *Measuring performance one product at a time.* Retrieved November 12, 2012, from http://www.apple.com/environment/*reports/*

Apple. (2013). *We believe in accountability—for our suppliers and for ourselves.* Retrieved June 3, 2013, from http://www.apple.com/*supplierresponsibility/*accountability.html

Automotive Industry Action Group Initiative. (2013). *AIAG—corporate responsibility.* Retrieved May 2, 2013, from http://www.aiag.org/staticcontent/committees/index.cfm#.UWWM-Rnfv0Q

Avorn, J. L., Barrett, J. F., Davey, P. G., McEwen, S. A., O'Brien, T. F., & Levy, S. B. (2001). *Antibiotic resistance: Synthesis of recommendations by expert policy groups.* Retrieved February 2, 2012, from http://whqlibdoc.who.int/hq/2001/WHO_CDS_CSR_DRS_2001.10.pdf

Barringer, F. (2012, February 2). Three states to require insurers to disclose climate change response plans. *New York Times.*

Battles, S. J. (2000). *Trends in residential air-conditioning usage from 1978 to 1997.* Retrieved April 5, 2012, from http://www.eia.gov/emeu/consumptionbriefs/recs/actrends/recs_ac_trends.html

Beachy, B., & Zorn, J. (2012). 21st Century GDP: National Indicators for a New Era (Masters in Public Policy, Trans.). In Harvard Kenndy School of Government (Ed.), *Policy analysis exercises* (p. 5). Cambridge, MA: Harvard University.

Blasi, G., & Doherty, J. (2010). *California employment discrimination law and its enforcement: The fair employment and housing act at 50.* Los Angeles, CA:

UCLA-RAND Center for Law and Public Policy, Forthcoming, Paper no. 10-06.

BMW Group. (2009). *Vehicle recycling. Focusing on sustainability.* Retrieved February 16, 2012, from BMW Report: http://www.bmwgroup.com/ publikationen/e/2009/pdf/2009_Vehicle_Recycling_Focusing_on_ Sustainability.pdf

Botzas, S. (2011). *Put your green deeds to the test.* Retrieved January 9, 2011, from The Sunday Times: http://www.thesundaytimes.co.uk/sto/public/ Appointments/article502258.ece

Boyd, B., Henning, N., Reyna, E., Wang, D. E., & Welch, M. D. (2009). *Hybrid organizations : New business models for environmental leadership.* Sheffield, UK: Greenleaf.

Brown, A. (2013, February). Shareholder support for CSR proposals doubles since 2005. *IR Magazine.*

Brundtland, G. (1987). *Our common future.* Oxford, UK: Oxford University Press.

Burman, L. (2011). *Should we care about rising income inequality?* Retrieved July 6, 2011, from Forbes: http://www.forbes.com/sites/leonardburman/2011/07/06/ should-we-care-about-rising-income-inequality/

CAN International. (n.d.). *About CAN Home Page.* Retrieved May 2, 2013, from http://www.climatenetwork.org/about/about-can

Casey, T. (2012). *Coke pulls the plug on anti-climate change ALEC lobby.* Retrieved April 6, 2012, from http://www.triplepundit.com/2012/04/coca-cola-compan-pulls-funding-from-alec/

CDP. (2013a). *CDP cities program.* Retrieved April 17, 2013, from https://www. cdproject.net/en-US/Programmes/Pages/cdp-cities.aspx

CDP. (2013b). *CDP suppply chain program.* Retrieved April 17, 2013, from https://www.cdproject.net/en-US/Programmes/Pages/CDP-Supply-Chain. aspx

CDP. (2013c). *Climate change program.* Retrieved April 17, 2013, from https:// www.cdproject.net/en-US/Programmes/Pages/CDP-Investors.aspx

CDP. (2013d). *CDP water program.* Retrieved April 17, 2013, from https://www. cdproject.net/en-US/Programmes/Pages/cdp-water-disclosure.aspx

CDP–What we do. (2013a). *CDP investor initiatives.* Retrieved April 17, 2013, from https://www.cdproject.net/en-US/WhatWeDo/Pages/investors.aspx

CDP–What we do. (2013b). *Government & policymakers.* Retrieved April 17, 2013, from https://www.cdproject.net/en-US/WhatWeDo/Pages/ government-policymakers.aspx

Center for Constitutional Rights. (2009). *The case against Shell.* Retrieved April 17, 2012, from Factsheet: http://ccrjustice.org/learn-more/faqs/ factsheet%3A-case-against-shell-0

Checkland, P. (1999). *Systems thinking, systems practice.* Chichester, EN: Wiley.

Chouinard, Y., & Stanley, V. (2012). *The responsible company.* Ventura, CA: Patagonia Books.

Cone Communications. (2012). *Cone green gap trend tracker.* Boston, MA: Cone Communications.

Crooks, R. (2009). *Mint map: Resource consumption by country.* Retrieved April 17, 2013, from MintLife Blog: www.mint.com/blog/trends/mint-map-resource-consumption-by-country/

Culp, K., Brooks, M., Rupe, K., & Zwerling, C. S. (2008). Traumatic injury rates in meatpacking plant workers. *Journal of Agromedicine 13*(1), 7–16.

Currie, A. (2012). China releases rare earth white paper. In R. E. I. News (Ed.), *Rare earth investing news.* Vancouver, BC: Investing News Network.

Dawson, J. (2004). Business leaders urged to heed global warming science. *Physics Today 57*(10), 37.

Deloitte (2013). *Collective responses to rising water challenges. CDP global water report 2012, 1.0.* Retrieved April 17, 2013, from https://www.cdproject.net/CDPResults/CDP-Water-Disclosure-Global-Report-2012.pdf

Di Tella, R., & MacCulloch, R. (2006). Some uses of happiness data in economics. *Journal of Economic Perspectives 20*, 25–46.

Diamond, J. M. (2006). *The third chimpanzee: the evolution and future of the human animal.* New York, NY: HarperPerennial.

Diener, E., Ng, W., Harter, J., & Arora, R. (2010). Wealth and happiness across the world: Material prosperity predicts life evaluation, whereas psychosocial prosperity predicts positive feeling. *Journal of Personality and Social Psychology 99*(1), 52–61.

Dietz, T., Rosa, E. A., & York, R. (2009). Environmentally efficient well-being: Rethinking sustainability as the relationship between human well-being and environmental impacts. *Human Ecology Review 16*(1), 114–123.

Earth Hour. (n.d.). *Earth hour.* Retrieved May 2, 2013, from http://www.earthhour.org/page/media-centre/earth-hour-history

Edwards, A. R. (2010). *Thriving beyond sustainability: Pathways to a resilient society.* Gabriola Island, B.C.: New Society Publishers.

Ehrenfeld, J., & Hoffman, A. J. (2013). *Flourishing: A frank conversation about sustainability.* Stanford, CA: Stanford Business Books.

Elkington, J. (2005). Enter the triple bottom line. In A. Henriques & J. Richardson (Eds.), *The triple bottom line: Does it all add up?* (pp. 1–14). London, GB: Earthscan.

Entine, J. (2012). *Campbell's big fat green BPA lie — and the sustainability activists who enabled it.* Retrieved September 18, 2012, from http://paradigmsanddemographics.blogspot.in/2012/09/campbells-big-fat-green-bpa-lie-and.html

Entrekin, S., Evans-White, M., Johnson, B., & Hagenbuch, E. (2011). Rapid expansion of natural gas development poses a threat to surface waters. *Frontiers in Ecology and the Environment 9*(9), 503–511.

EPA. (2011). *Municipal solid waste generation, recycling, and disposal in the united states: facts and figures for 2010.* Retrieved April 17, 2012, from http://www.epa.gov/osw/nonhaz/municipal/pubs/msw_2010_rev_factsheet.pdf

Ernst, & Young. (2013). 2013 *Six growing trends in corporate sustainability.* Retrieved April 7, 2013, from http://www.ey.com/Publication/vwLUAssets/Six_growing_trends_in_corporate_sustainability_2013/$FILE/Six_growing_trends_in_corporate_sustainability_2013.pdf

Exxon Mobil. (2012). Taking on the world's toughest energy challenges, 2012 10-K Annual Report. (Report Number P.61).

Federal Register. (2009). *U.S. Government Printing Office home page. Volume 74 Issue 194 (Thursday, October 8, 2009).* Retrieved May 2, 2013, from http://www.gpo.gov/fdsys/pkg/FR-2009-10-08/html/E9-24518.htm

Fellow, A. (2013). *Investors demand climate-risk disclosure in 2013 proxies - Bloomberg.* Retrieved May 2, 2013, from Bloomberg–Business, Financial & Economic News, Stock Quotes: http://www.bloomberg.com/news/2013-02-25/investors-demand-climate-risk-disclosure-in-2013-proxies.html

Financial Times. (2012). *FT 500. World business, finance, and political news from the Financial Times–FT.com.* Retrieved May 2, 2013, from http://www.ft.com/companies/ft500

Fox, E. J. (Producer). (2013). *Bangladesh factory collapse kills at least 160, reviving safety questions.* Retrieved April, 24, 2013, from CNNMoney: http://money.cnn.com/2013/04/24/news/companies/bangladesh-factory-collapse/index.html

Freeman, R. E. (1984). *Strategic management: A stakeholder approach.* Boston, MA: Pitman.

Gallucci, M. (2012, September 6). Major corporations quietly reducing emissions—and saving money. *InsideClimate News.*

Geary, D. C. (2005). *The origin of mind: Evolution of brain, cognition, and general intelligence.* Washington, D.C: American Psychological Association.

Global Reporting Initative (2013). *Trends in external assurance of sustainabilityreports: Spotlight on the USA.* Retrieved May 2, 2013, from Global Reporting Initiative: http://www.ga-institute.com/research-reports/gri-focal-point-us- assurance-trends-study.html

Gressel, J. (1999). Tandem constructs: preventing the rise of superweeds. *Trends in biotechnology,* 17(9), 361–366.

GRI. (2013). *About global reporting initiative.* Retrieved May 2, 2013, from https://www.globalreporting.org/Information/about-gri/Pages/default.aspx

Gupta, V., & Sengupta, R. (2007). *Monsanto's 'Roundup Ready' alfalfa controversy* (pp. 12). Hyderabad, IN: ICMR Center for Management Research.

Hart, S. L. (2008). Beyond greening: Strategies for a sustainable world. In Harvard Business Review (Ed.), *Harvard Business Review paperback series*

(Vol. Harvard Business Review on Profiting from Green Business, pp. 105–130). Boston, MA: Harvard Business School Publishing Corporation.

Harter, J. K., Schmidt, F. L., Asplund, J. W., Killham, E. A., & Agrawal, S. (2010). Causal impact of employee work perceptions on the bottom line of organizations. *Perspectives on Psychological Science* 5(4), 378–389.

Hawken, P. (1993). *The ecology of commerce: A declaration of sustainability* (Rev. ed.). New York, NY: Harper Business.

Hawken, P., Lovins, A. B., & Lovins, L. H. (2000). *Natural capitalism: Creating the next industrial revolution* (1st Back Bay pbk. ed.). Boston, MA: Little, Brown and Co.

Hitchcock, D., & Willard, M. (2008). *The step-by-step guide to sustainability planning.* Sterling, VA: Earthscan.

Hoffman, A. J., & Bazerman, M. H. (2007). Changing practice on sustainability: understanding and overcoming the organizational and psychological barriers to action. In S. Sharma, M. Starik & B. Husted (Eds.), *Organizations and the sustainability mosaic* (pp. 84–105). Northampton, MA: Edward Elgar Publishing.

ICLEI Global. (n.d.). *ICLEI global cities network.* Retrieved May 2, 2013, from http://www.iclei.org/

International Council of Mining and Metals. (2012). *Mining's Contributions to Sustainable Development—An Overview.* London, UK: Trends in Mining and Metals.

International Insurance Society. (2013). *Sustainability and innovation to capitalize on global trends.* Retrieved March 3, 2013, from http://www.iisonline.org/forum/market-trends/

IPCC. (2006). *Task force on national greenhouse gas inventories.* Retrieved May 2, 2013, from http://www.ipcc-nggip.iges.or.jp/public/2006gl/

IPCC. (2007). Summary for policymakers. In B. Metz, O. R. Davidson, P. R. Bosch, R. Dave & L. A. Meyer (Eds.), *In: Climate change 2007: Mitigation. contribution of working group III to the fourth assessment report of the intergovernmental panel on climate change.* Cambridge, MA: Intergovernmental Panel on Climate Change.

Jegarajah, S. (Producer). (2012). *Conflict minerals in your mobile—Why Congo's war matters.* Retrieved January 6, 2013, from CNBC Asia-Pacific News: http://www.cnbc.com/id/49961559

Jensen, M. C., & Meckling, W. H. (1994) The nature of man (revised July 1997). *Journal of Applied Finance* 7(2), 4–19.

Jensen, M. C., & Meckling, W. H. (1976). Theory of the firm: Managerial behavior, agency costs and ownership structure. *Journal of Financial Economics* 3(4), 305–360.

Jørgensen, A., Finkbeiner, M., Jørgensen, M. S., & Hauschild, M. Z. (2010). Defining the baseline in social life cycle assessment. *The International Journal of Life Cycle Assessment* 15(4), 376–384.

Kahneman, D. (2011). *Thinking, fast and slow*. New York, NY: Farrar, Straus and Giroux.

Kanaplue, S., & Berman, R. (2012). *Project: Green products–a sustainability assessment: Artificial christmas tree vs. Natural christmas tree*. Retrieved November 14, 2012, from http://www.greendesignetc.net/GreenProducts_12/GreenProduct_kanaplue_Steven_Paper.pdf

Kessler, J., & Ellis, B. (2011). *Bank of America axes $5 debit card fee*. Retrieved April 7, 2012, from http://money.cnn.com/2011/11/01/pf/bank_of_america_debit_fee/index.htm

Kiron, D. (2012). *Get ready: Mandated integrated reporting is the future of corporate reporting*. Retrieved March 13, 2012, from MIT Sloan Management Review: http://sloanreview.mit.edu/feature/get-ready-mandated-integrated-reporting-is-the-future-of-corporate-reporting/

Kiron, D., Kruschwitz, N., Reeves, M., & Goh, E. (2013). The benefits of sustainability-driven innovation. *MIT Sloan Management Review 54*(2), 69–73.

Klugman, J. (2011). Summary: Human development report 2011 - Sustainability and equity, a better future for all *Human development report* (pp. 26). New York, NY: The UNDP Human Development Report Office.

KPMG Advisory N. V., Global Reporting Initiative, & Centre for Corporate Governance in Africa. (2013). *Carrots and Sticks: Sustainability reporting policies worldwide – Today's best practice, tomorrow's trends*. Amsterdam, NL: Global Reporting Initiative.

KPMG Global. (2013). *Corporate sustainability: A progress report*. Retrieved May 2, 2013, from http://www.kpmg.com/global/en/issuesandinsights/articlespublications/pages/corporate-sustainability.aspx

KPMG International in cooperation with Economist Intelligence Unit. (2011a). *Corporate sustainability: A progress report* Retrieved April 2011, from http://www.kpmg.com/Global/en/IssuesAndInsights/ArticlesPublications/Pages/corporate-sustainability.aspx

KPMG International in cooperation with Economist Intelligence Unit. (2011b). *KPMG International survey of corporate responsibility reporting 2011*. Retrieved November 2011, from http://www.kpmg.com/PT/pt/IssuesAndInsights/Documents/corporate-responsibility2011.pdf

Kurapatskie, B., & Darnall, N. (2012). *Are some corporate sustainability activities associated with greater financial payoffs?* Retrieved July 26, 2012, from Business Strategy and the Environment, Forthcoming at SSRN: http://ssrn.com/abstract=2118261

Liebenthal, A., Michelitsch, R., & Tarazona, E. (2005). *Extractive industries and sustainable development: An evaluation of world bank group experience* (O. E. Department, Trans.) (p. 1). Washington, D.C.: World Bank Operations Evaluation Department.

Ling, A., Forrest, S., Fox, M., & Feilhauer, S. (2007). *Introducing GS SUSTAIN*. Portland, ME: The Goldman Sachs Group, Inc. Global Investment Research.

Lovins, A. B., Lovins, L. H., & Hawken, P. (1999). A road map for natural capitalism. *Journal of Business Administration and Policy Analysis, 27–29*.

Lubber, M. (2012). *A tipping point on sustainability disclosure in Rio? Forbes*. Retrieved June 6, 2012, from http://www.forbes.com/sites/mindylubber/2012/06/19/a-tipping-point-on-sustainability-disclosure-in-rio/

Luhby, T. (2011). *Global income inequality: Where the U.S. ranks. Money, November 8. 2011*. Retrieved February 20, 2012, from http://money.cnn.com/2011/11/08/news/economy/global_income_inequality/index.htm

Macarthur E. (2013). *The circular model: An overview*. Retrieved July 8, 2013, from http://www.ellenmacarthurfoundation.org/circular-economy/circular-economy/the-circular-model-an-overview

Macfarlane, J. (2013). *Industry process map*. Retrieved from http://doctorjane.org/industry-process-maps/

Mahler, D., Barker, J., Belsand, L., & Schulz, O. (2009). *Green winners: The performance of sustainability-focused companies during the financial crisis*. Chicago, IL: A.T Kearney, Inc.

Mair, V. H. (2005). *Danger + opportunity ≠ crisis: How a misunderstanding about Chinese characters has led many astray*. Retrieved February 12, 2012, from http://pinyin.info/chinese/crisis.html

Makower, J. (2005). *Ecomagination: Inside GE's power play.*Retrieved May 2, 2013, from GreenBiz.com: http://www.greenbiz.com/blog/2005/05/10/ecomagination-inside-ges-power-play

Makower, J. (2012). *Newsweek's 2012 green rankings: This time it's serious*. Retrieved May 2, 2013, from GreenBiz.com: http://www.greenbiz.com/blog/2012/10/20/newsweeks-2012-green-rankings-time-its-serious

Makower, J. (2013). *State of Green Business Report 2013.*. Retrieved May 2, 2013, from GreenBiz.com: http://www.greenbiz.com/research/report/2013/02/state-green-business-report-2013

Manzini, R., Noci, G., Ostinelli, M., & Pizzurno, E. (2004). Assessing environmental product declaration opportunities: A reference framework. *Business Strategy and Development 15*(2), 118–134.

MarineBio. (2013). *Ocean resources–marinebio conservation society*. Retrieved October 23, 2013, from http://marinebio.org/oceans/ocean-resources.asp

Martin R. (2010, January). The age of customer capitalism. *Harvard Business Review* [serial online], pp. 58–65.

Matsumura, E. M., Prakash, R., & Vera-Muñoz, S. C. (2011). *Voluntary disclosures and the firm-value effects of carbon emissions*. Retrieved April 26, 2011, from http://business.nd.edu/news_and_events/news_articles_article.aspx?id=9054

McCulloch, J. (2009). Counting the cost: Gold mining and occupational disease in contemporary South Africa. *African Affairs 108*(431), 221.

McDonough, W., & Braungart, M. (2013). *The upcycle: Beyond sustainability - designing for abundance.* New York, NY: North Point Press, a division of Farrar, Straus and Giroux.

McGranahan, G., Balk, D., & Anderson, B. (2007). The rising tide: assessing the risks of climate change and human settlements in low elevation costal zones. *Environment and Urbanization 19*(1), 17–37.

McIntire, L. (2011). *Why UPS's sky-high carbon disclosure project score wasn't an accident.* Retrieved May 2, 2013, from http://blog.ups.com/2011/10/05/why-upss-sky-high-carbon-disclosure-project-score-wasn%E2%80%99t-an-accident/

McLamb, E. (2011). *Impact of the industrial revolution.* Retrieved May 2, 2013, from Ecology Global Network: http://www.ecology.com/2011/09/18/ecological-impact-industrial-revolution/

Meadows, D. H. (1972). *The Limits to growth; A report for the Club of Rome's project on the predicament of mankind.* New York, NY: Universe Books.

Meadows, D. H. (1999). Leverage points: Places to intervene in a system. Hartland, VT: The Sustainability Institute and Chapter 1. In P. M. Senge, (Ed.) (2006). *The fifth discipline.* London, UK: Random House Business.

Meadows, D. H. (2008). *Thinking in systems: A primer.* White River Junction, VT: Chelsea Green Publication.

Meadows, D. H., Meadows, D. L., Randers, J., & Behrens III, W. W. (1972). *The Limits to growth; A report for the Club of Rome's project on the predicament of mankind.* New York, NY: Universe Books.

Merck. (2011). *Corporate responsibility report 2011.* Retrieved May 2, 2012, from http://www.merckresponsibility.com/focus-areas/environmental-sustainability/home.html

Miller, N. G., Pogue, D., Gough, Q. D., & Davis, S. M. (2009). Green buildings and productivity. *The Journal of Sustainable Real Estate 1*(1), 65–89.

Mintzberg, H., Robert, S., & Basu, K. (2002). Beyond selfishness. *MIT Sloan Management Review 44*(1), 67.

Molina, B. (2012). *In U.S. building industry, is it too easy to be green?* Retrieved May 2, 2013, from USA TODAY: http://www.usatoday.com/story/news/nation/2012/10/24/green-building-leed-certification/1650517/

Morello, L., & ClimateWire. (2010). *Phytoplankton population drops 40 percent since 1950.* Retrieved July 29, 2010, from Scientific American website: http://www.scientificamerican.com/article.cfm?id=phytoplankton-population

Motavalli, J. (2013, March 13). Automakers work to achieve zero-waste goals. *New York Times.*

Munich RE. (2013). 2012 *Natural catastrophe statistics year in review*. Retrieved May 2, 2013, from Munich RE: www.munichre.com/en/media_relations/press_releases/2013/2013_01_03_press_release.aspx

Musco Family Olive Co.–Environmental Initiatives. (2013). *Musco family olive co.—All about olives–Olive recipes and much more*. Retrieved May 2, 2013, from http://www.olives.com/environment.html

National Aeronautics and Space Administration. *The current and future consequences of global change*. Retrieved from http://climate.nasa.gov/effects/

Nattrass, B. F., & Altomare, M. (1999). *The natural step for business: Wealth, ecology, and the evolutionary corporation* (p. 42.). Gabriola Island, BC: New Society Publishers.

Natural Capital Solutions. (2012). *Sustainability pays: Studies the prove the business case for sustainability*. Retrieved May 2, 2013, from Natural Capital Solutions, 1.0.: http://www.natcapsolutions.org/businesscasereports.pdf

Nestle. (2012). *Nestle sustainability report 2012*. Retrieved May 2, 2013, from http://www.nestle.com/csv/environmental-sustainability/transport-distribution

Neuman, W. (2011, November 24). A question of fairness. *The New York Times*.

New Belgium Brewing. (2013). *Sustainability section of website*. Retrieved May 3, 2013, from http://www.newbelgium.com/sustainability.aspx

Newmont. (2012a). *2012 Annual report*. Retrieved May 3, 2013, from http://www.newmont.com/sites/default/files/Newmont_2012_annual_report.pdf

Newmont. (2012b). *Beyond the mine. 2012 Sustainability Report*. Retrieved May 3, 2013, from http://www.beyondthemine.com/2012/

OECD Forum on Tackling Inequality (2011). *Growing income inequality in OECD countries: What drives it and how can policy tackle it? May, 2011, Paris: OECD*. Retrieved February 1, 2012, from www.oecd.org/dataoecd/32/20/47723414.pdf

Oikos International. (2013). *www.oikos-international.org - oikos International*. Retrieved May 2, 2013, from http://www.oikos-international.org/academic/case-collection/

Oreskes, N. (2004). Beyond the ivory tower: The scientific consensus on climate change. *Science 306*(5702), 1686.

Orlando Business Journal. (2011). *Peabody Orlando launches composting program*. Retrieved September 22, 2011, from: http://www.bizjournals.com/orlando/news/2011/09/22/peabody-orlando-launches-composting.html

Oswald, A. J., Proto, E., & Sgroi, D. (2009). *Happiness and productivity*, IZA discussion papers, No. 4645.

Patagonia. (2012). *Changing ports pays dividends*. Retrieved May 2, 2013, from http://www.patagonia.com/us/patagonia.go?assetid=79365

Porter, M. E., & Kramer, M. R. (2011). Creating shared value. [Article]. *Harvard Business Review 89*(1/2), 62–77.

Post, J. E., & Altman, B. W. (1994). Managing the environmental change process: barriers and opportunities. *Journal of Organizational Change Management 7*(4), p. 64–66.

Rain Forest Action. (2012). *Rainforest Action Network Annual Report 2012.* http://ran.org/sites/default/files/ran_ar_2012_vlow.pdf

Rajaram, D. (2011, May). Making the business case for sustainability. *Harvard Business Report Blog.*

Raloff, J. (2012, November 17). Extremely bad weather. *Science News 182,* p. 22.

Reuben, A. (2012). *The Higg Index for sustainable apparel* (Case Study). Retrieved September 18, 2013, from Yale Center for Environmental Law & Policy http://epi.yale.edu/indicators/indicator-case-studies/reports/higg-index-sustainable-apparel

Revkin, A. C. (2009). *Campaign against emissions picks number.* Retrieved May 2, 2013, from The New York Times - Breaking News, World News & Multimedia: http://www.nytimes.com/2009/10/25/science/earth/25threefifty.html?_r=0

Ricketts, C. (2010). *GE pumps $10B more into green technology R&D | VentureBeat.* Retrieved May 2, 2013, from VentureBeat: http://venturebeat.com/2010/06/24/ge-pumps-10b-more-into-green-technology-rd/

RobecoSAM. (2007). *Sustainability indexes–home.* Retrieved May 2, 2013, from http://www.sustainability-indices.com/sustainability-assessment/corporate-sustainability.jsp

Roston, E. (2012). *Non-financial data is material: The sustainability paradox.* Retrieved April 15, 2012, from http://www.businessweek.com/news/2012-04-13/non-financial-data-is-material-the-sustainability-paradox.

SAIC. (2006). *Life cycle assessment: Principles and practice, EPA/600/R-06/060.* Cincinnati, OH: US Environmental Protection Agency/Office of Research & Development.

Salisbury, P. (2011). *Behind the brand: McDonald's.* Retrieved April 7, 2012, from http://www.theecologist.org/green_green_living/behind_the_label/941743/behind_the_brand_mcdonalds.html

Sarkar, M. (2011). How American homes vary by the year they were built. *Housing and economic statistics* (Vol. Working Paper No. 2011-18). Washington, DC: U.S. Census Bureau.

Saunders, T., & McGovern, L. (1993). *The bottom line of green is black* (p. 107–111). San Francisco, CA: Harper.

Savitz, A. W., & Weber, K. (2006). *The triple bottom line: How today's best-run companies are achieving economic, social, and environmental success-and how you can too.* San Francisco, CA: Jossey-Bass.

Senge, P. M. (2006). *The fifth discipline.* London, GB: Random House Business.

Senge, P. M. (2008). *The necessary revolution: How individuals and organizations are working together to create a sustainable world.* New York, NY: Doubleday.

Servaes, H., & Tamayo, A. M. (2012). *The impact of corporate social responsibility on firm value: The role of customer awareness.* Retrieved July 1, 2012, from SSRN: http://ssrn.com/abstract=2116265

Shaw, B., & Mozley, C. (1962). *Man and superman.* New York, NY: Ltd. Editions Club.

Shontell, A. (2011, March 24). Why UPS is so efficient: Our trucks never turn left. *Business Insider.*

Singer, F. S., & Avery, D. T. (2005). *The physical evidence of earth's unstoppable 1,500-year climate cycle.* Dallas, TX: National Center for Policy Analysis.

SME's set their sights on Sustainability. (2011). *SME's set their sights on sustainability.* Retrieved May 1, 2013, from www.aicpa.org/interestareas/businessindustryandgovernment/resources/sustainability/downloadabledocuments/sustainability_case_studies_final%20pdf.pdf

Social Progress Index. (2013). *Measuring national process.* Retrieved September 17, 2013, from http://www.socialprogressimperative.org/data/spi

Sustain Ability. (2007). *Stakeholder resource: A concise and dense resource to learn more about the principles and practices for successful stakeholder engagement.* Retrieved October 2007, from, http://www.csr-weltweit.de/uploads/tx_jpdownloads/SustainAbility_Practices_and_Principles_for_successful_stakeholder_engagement.pdf

Starbuck, S., Brockett, A., Gilbert, B., LeBlanc, B., Naumoff, P., & Walker, C. (2012). *Six growing trends in corporate sustainability: An Ernst & Young survey in cooperation with GreenBiz Group.* Retrieved February 6, 2012, from http://www.greenbiz.com/research/report/2012/03/01/six-growing-trends-corporate-sustainability

Stead, W. E., & Stead, J. G. (1994). Can humankind change the economic myth? Paradigm shifts necessary for ecologically sustainable business. *Journal of Organizational Change Management, 7*(4), p. 15–25.

Stiglitz, J. (2012). Is mercantilism doomed to fail? *Paradigm lost: Rethinking economics and politics.* New York, NY: Institute for New Economic Thinking.

Sullivan, N., & Schiafo, R. (2005). *Talking green, acting dirty.* Retrieved May 2, 2013, from The New York Times–breaking news, world news & multimedia: http://www.nytimes.com/2005/06/12/opinion/nyregionopinions/12WEsullivan.html

Sulzberger, A. G. (2011, April 14). States look to ban efforts to reveal farm abuse. *The New York Times.*

Sustainability Accounting Standards Board. (2013). *Sustainability accounting standards board.* Retrieved May 2, 2013, from http://www.sasb.org/

Sustainability Disclosure Database. (2013). *Sustainability disclosure database–home.* Retrieved May 2, 2013, from http://database.globalreporting.org/

Sustainable Growth 2012. (2013). *Download the 2012 report.* Retrieved September 10, 2013, from http://www.gecitizenship.com/2012-report/download-the-2012-report/

Sweetlove, L. (2011). *Number of species on earth tagged at 8.7 million.* Retrieved August 8, 2012, from Nature: http://www.nature.com/news/2011/110823/full/news.2011.498.html

Swibel, M. (2007). *The 50 top microfinance institutions.* Retrieved December 20, 2007, from Forbes: http://www.forbes.com/2007/12/20/microfinance-philanthropy-credit-biz-cz_ms_1220microfinance_table.html

Talberth, J. (2012). *Measuring genuine progress: Towards global consensus on a headline indicator for the new economy.* Retrieved June 2012, from http://genuineprogress.net/wp-content/uploads/2013/01/Measuring-Genuine-Progress-Final.pdf

The Climate Conservatory. (2008). *The carbon footprint of Fat Tire Amber Ale.* Retrieved June, 2012 from http://www.newbelgium.com/Files/the-carbon-footprint-of-fat-tire-amber-ale-2008-public-dist-rfs.pdf

The Project on Scientific Knowledge and Public Policy. (2012). *Diacetyl / Popcorn Workers Lung.* Retrieved June 15, 2012, from Defending Science: http://defendingscience.org/case-studies/diacetyl-background

The World Bank. (2012). *Turn down the heat. Why a 4 degree centrigrade warmer world must be avoided.* Retrieved April 17, 2013, from http://climatechange.worldbank.org/sites/default/files/Turn_Down_the_heat_Why_a_4_degree_centrigrade_warmer_world_must_be_avoided.pdf

Thomas, K. (2012). *Johnson & Johnson to remove formaldehyde from product..* Retrieved August 15, 2012, from The New York Times: http://www.nytimes.com/2012/08/16/business/johnson-johnson-to-remove-formaldehyde-from-products.html?_r=1

United National Secretary-General's Eminent Persons Group on the Least Developed Countries (2011). *Compact for inclusive growth and prosperity. March, 2011.* Retrieved February 20, 2012, from http://www.un.org/wcm/webdav/site/ldc/shared/EPG_Report_ENGLISH_w_v2.pdf

United Nations Environmental Program. (n.d.). Retrieved May 2, 2013, from http://www.unep.org/Documents.Multilingual/Default.Print.asp?DocumentID=52&ArticleID=60

United Nations Global Compact. (2013). *Global corporate sustainability report 2013.* New York, NY: United National Global Compact.

United Nations Human Rights. (2012). *Report of the special rapporteur on the rights of indigenous peoples.* Retrieved May 2, 2013, from http://www.ohchr.org/EN/Issues/IPeoples/SRIndigenousPeoples/Pages/AnnualReports.aspx

United States Environmental Protection Agency. (2013). *Future climate change.* Retrieved February 6, 2012, from http://www.epa.gov/climatechange/science/future.html

UPS. (2012, July 31). UPS sustainability report hits a plus mark for transparency. *UPS Press Room.*

US EPA. (2013). *Climate change: Basic information.* Retrieved May 2, 2013, from http://www.epa.gov/climatechange/basics/

Van Nood et al. (2013). Duodenal infusion of donor feces for recurrent Clostridium Difficile. *The New England Journal of Medicine 368*(5), 407–415.

Visser, W. (2011). *The age of responsibility: CSR 2.0 and the new DNA of business.* Hoboken, NJ: John Wiley and Sons.

Visser, W., & University of Cambridge Programme for Sustainability Leadership. (2009). *The top 50 sustainability books.* Sheffield, UK: Greenleaf Publishing.

Von Grebmer, K., Fritschel, H., Nestorova, B., Olofinbiyi, T., Pandya-Lorch, R., & World Meteorological Organization. (2009). *Experts agree on a universal drought index to cope with climate risks.* Retrieved December 15, 2009, from A universal drought index WMO website: http://www.wmo.int/pages/mediacentre/press_releases/pr_872_en.html

Walmart. (2013a). *Global responsibility report.* Retrieved May 2, 2013, from http://corporate.walmart.com/microsites/global-responsibility-report-2013/

Walmart. (2013b). *Walmart corporate sustainability–support home page.* Retrieved May 2, 2013, from http://www.walmartsustainabilityhub.com/

Walmart. (2013c). Walmart Stores, Inc. 2013 Form 10-K. P. 40. Retrieved April 7, 2013.

Warner, K., Timme, W., Lowell, B., & Hirshfield, M. (2012). *Widespread seafood fraud found in L.A.* Washington, DC: Oceana.

Waste Management. (2012). *We are all about solutions.* Retrieved September 2012, from Waste Management 2012 Annual Report: http://investors.wm.com/phoenix.zhtml?c=119743&p=irol-reportsannual

Watkins, K. (2008). Human development report 2007/2008: Fighting climate change: Human solidarity in a divided world: *Human Development Report Summary.* New York, NY: United Nations Development Program.

Webb, T. (2012). *The future of integrated sustainability reporting.* Retrieved May 2, 2013, from GreenBiz.com: http://www.greenbiz.com/news/2012/11/26/future-integrated-sustainability-reporting?page=0%2C1

Whitea, C. M., Strazisara, B. R., Granitea, E. J., Hoffmana, J. S., & Pennlinea, H. W. (2012). Separation and capture of CO_2 from large stationary sources and sequestration in geological formations—Coalbeds and deep saline aquifers. *Journal of the Air & Waste Management Association 53*(6), 645–715.

WikiInvest. (2007). *Proposal 5: To require a sustainability report. Comcast shareholder resolutions.* Retrieved May 2, 2013, from www.wikinvest.com/stock/Comcast_%28CMCSA%29/Proposal_Require_Sustainability_Report

Willard, B. (2005). *The next sustainability wave: Building boardroom buy-in.* Gabriola Island, BC: New Society Publishers.

Willard, B. (2009). *The sustainability champion's guidebook: How to transform your company*. Gabriola Island, BC: New Society Publishers.

Wimmer, W., Züst, R., & Lee, K. M. (2004). *Ecodesign implementation: A systematic guidance on integrating environmental considerations into product development* (Vol. 6). Norwell, MA: Kluwer Academic Pub.

Winston, A. (2010). *The most powerful green NGO you've never heard of*. Retrieved October 5, 2010, from Harvard Business Review: http://blogs.hbr.org/winston/2010/10/the-most-powerful-green-ngo.html

Wolpert, S. (2011). *Species extinction rates have been overestimated, new study claims; However, researchers say, global extinction crisis remains very serious.* Retrieved May 6, 2012, from UCLA Newsroom: newsroom.ucla.edu/portal/ucla/species-extinction-rates-have-203632.aspx

World Economic Forum and Accenture. (2012). *More with less: Scaling sustainable consumption and resource efficiency.* Retrieved May 2013, from http://www3.weforum.org/docs/IP/2012/CO/WEF_CO_ScalingSustainable ConsumptionResourceEfficiency_ExecutiveSummary_2012.pdf

Worldwatch Institute. (2012). *Fact sheet: The impacts of weather and climate change.* Retrieved May 1, 2012, from http://www.worldwatch.org/node/1779

Yates, J. L. (2012). Abundance on trial: The cultural significance of sustainability. *The Hedgehog Review 14*(2).

Yohannes, Y. (2008). *Global hunger index: The challenge of hunger 2008.* Washington, DC: International Food Policy Research Institute.

Zimmer, C. (2010). *A looming oxygen crisis and its impact on world's oceans..* Retrieved August 5, 2010, from Environment 360 website: http://e360. yale.edu/feature/a_looming_oxygen_crisis_and_its_impact_on_worlds_ oceans/2301/

Zimmer, C. (2012, June 16). Tending the body's microbial garden. *The New York Times.*

Index

OTHER TITLES IN OUR ENVIRONMENTAL AND SOCIAL SUSTAINABILITY FOR BUSINESS ADVANTAGE COLLECTION

Chris Laszlo, Case Weatherhead School of Management
and Robert Sroufe, Duquesne University

- *Strategy Making in Nonprofit Organizations: A Model and Case Studies* by Jyoti Bachani
- *IT Sustainability for Business Advantage* by Brian Moore
- *A Primer on Sustainability: In the Business Environment* by Ronald M. Whitfield and Jeanne McNett
- *Change Management for Sustainability* by Houng Ha Houng

Announcing the Business Expert Press Digital Library

*Concise E-books Business Students Need
for Classroom and Research*

This book can also be purchased in an e-book collection by your library as
- a one-time purchase,
- that is owned forever,
- allows for simultaneous readers,
- has no restrictions on printing, and
- can be downloaded as PDFs from within the library community.

Our digital library collections are a great solution to beat the rising cost of textbooks. e-books can be loaded into their course management systems or onto student's e-book readers.

The **Business Expert Press** digital libraries are very affordable, with no obligation to buy in future years. For more information, please visit **www.businessexpertpress.com/librarians**. To set up a trial in the United States, please contact **Adam Chesler** at *adam.chesler@ businessexpertpress.com* for all other regions, contact **Nicole Lee** at *nicole.lee@igroupnet.com*.

www.ingramcontent.com/pod-product-compliance
Lightning Source LLC
Chambersburg PA
CBHW070915270326
41927CB00011B/2576